D0089771

Return of
the Bird Tribes

OTHER BOOKS BY KEN CAREY

The Starseed Transmissions
The Third Millennium
Vision
Flat Rock Journal

Return of
the Bird Tribes

Ken Carey

HarperSanFrancisco

A Division of HarperCollins*Publishers*

To Sherry

A TREE CLAUSE BOOK

HarperSanFrancisco and the author, in association with The Basic Foundation, a not-for-profit organization whose primary mission is reforestation, will facilitate the planting of two trees for every one tree used in the manufacture of this book.

Workshops and Seminars
For information on workshops and seminars by Ken Carey, write:
Starseed Seminars
Star Route, Box 70
Mountain View, Missouri 65548

FIRST HarperCollins PAPERBACK EDITION PUBLISHED IN 1991

Cover art, "Shaman's Vision," by Keith Powell, P.O. Box 788, Grand Coulee, WA 99133.

Library of Congress Cataloging-in-Publication Data
Carey, Ken.
Return of the bird tribes / Ken Carey.—1st HarperCollins pbk. ed.
p. cm.
ISBN 0–06–250188–7
1. Spirit writings. I. Title.
BF1301.C35 1991
133.9'3—dc20   90–56460

---

98 99 RRD-H 20 19 18 17 16 15 14

Acknowledgments

I would like to thank John Clancy and Marilee Talman for their insightful editorial suggestions during the final stages of bringing this book into form, Marilyn McGuire for her offerings too numerous to mention, Keith Powell for allowing us to use his exquisite painting "Shaman's Vision" on the book cover, Tadashi Kameda for leading the sweat lodge ceremony that was responsible for awakening such vital memories. I would like to thank the Karuk Tribe and the Earth Circle Association, Sun Bear and the Bear Tribe and all the people at Hopecorp. I also would like to extend a special expression of gratitude to all of the following for their inspiration, encouragement and support during various stages of the book's formation: Gary Heyde, Isabelle Ganter, Jeff Knaebel, Kate Houser-Boutherin, Lenore Carroll, John and Brenda Cutrell, Dr. John Harris, Joyce Yates, Clifford and Gwendolyn Vance, Lola Garcia, Darlene Ravin, Diana Elena, Tom and Eileen Carey, Marianne and Bill Schramm, Ronald and Meredyth Ross, Barbara Marx Hubbard, Dawson Church, Brenda Plowman, Joan Turner, Hiroshi and Naoko Haga, Ray Polichetti, Verna Morrell, Deborah Duda, Dr. Wayne Dyer, Laura Huxley, Virginia

ACKNOWLEDGMENTS

Satir, Fred McMinn, Mary LaMar, Robin Shipman, Susan Dresher, Gary Sycalik, Susan Verlander, Virginia Tibbetts, Pat Beals, Jean Houston, Sir George Trevelyan, Elisabeth Bottoni and Valarie Carey. Thank you all!

Contents

Introduction to
the 1995 Edition

With the acquisition of language, a vast range of perception is disregarded. A child soon comes to believe that if there is no word for something it does not exist. Sensory input that is not reinforced is deemphasized, until eventually the child accepts the cultural viewpoint, with all of its biases and underlying assumptions. In the process something valuable is lost.

The benefits of language are obvious. Without it, each new generation would repeat the mistakes and forego the learning of the past. Without it, our race could not advance. But the view of ourselves and our world that comes to us through language is but a thin sliver of the multidimensional reality we inhabit.

This larger reality has fascinated me since childhood. I have never been without at least some minimal awareness of it, and through the years there were always those odd moments when my mind would relax into a form of perception that brought me fully into its strange and curious domain. Often, this perception would come forward to help me establish a rapport with a plant or an animal deeper than what I found possible when viewing things from a cultural perspective. Although my language-oriented mind would later remember that these incidents occurred, since my experiences were nonverbal I rarely remembered much of their contents or detail.

It was not until one day in the winter of 1978 that I made my first attempt to verbalize what I experienced during one of these states. In bed with a cold and fever, my thoughts had drifted into a calm, meditative silence. Passing moments pooled together, relaxing into one. Distinctions between inside and outside began to dissolve. On a level beyond words, beyond the surface froth of passing events, on a level deep in the ocean of being, I experienced myself as energy, and my body, as an intertwining of sun and soil.

Was this the awareness of the newborn and the very young? Everything that touched my senses, every nuance of sound and light, every object in the room felt as intimate to me as the lungs through which I breathed, as inseparable, as personal. Insights did not come as flashes but as things I had always known, truths so obvious it was hard to believe I could have forgotten them. But I had . . . and might again. Could this awareness be expressed, articulated, recorded in words?

Language only limits perception when its terms are taken too seriously. Taken lightly, the attempt to find words to record our impressions can help us to perceive more fully. Verbalizing this awareness would challenge me to clarify my experience of it. Yet its very nature seemed to defy definition. To translate it into symbolic characters on sheets of paper would be like trying to channel a river through an hourglass. The very impossibility of the task intrigued me. I put a new ribbon in the typewriter.

During the eleven days that followed I wrote my first book, *The Starseed Transmissions*. I wrote that book and the companion volumes that followed to help me remember what I too often forget: that I inhabit a world far richer and more complex, far more potent and beautiful than the one superimposed by the interpretations of a species whose consciousness has yet to catch up with its capability.

While our technology has brought us to the moon, our thinking has remained clouded by assumptions formed in a primitive world of treetops and savanna. Our languages have conditioned us to see the earth through a filter of the past, a murky glass of mammalian wants and needs. We repeat the errors of history, forgetting that history is just a habit, and like any habit, it can be changed.

The four *Starseed* volumes—which conclude with *The Third Millennium*—combine to form a preliminary outline of a work in progress: a planetary work of biological art soon to become conscious of itself in a new and historically unprecedented way.

Certain chapters in these volumes speak of the past; others, of the future. All relate to the present. Like the pages of a blueprint, some chapters show the northern view of the work in progress, others, the southern, eastern, western. *Return of the Bird Tribes* provides examples of people whose passion for the truth revolutionized their societies: White Buffalo Calf Woman introducing the sacred pipe ceremony to the Sioux. The Great Peacemaker forging the Iroquois Confederacy. The overview of planetary evolution recorded in *Vision* compliments these historical insights with a first person singular narrative quite different from the plural used in *The Starseed Transmissions* and *The Third Millennium*. But differences define the spokes of a wheel, and without them our world would not turn.

To fully awaken the earth requires each of us: A blade of grass, a gull, a passing cloud, a squirrel, a dolphin, an oak tree. Spiritually, our race is superior to none of these.

Yet this truth is misapplied if it causes us to overlook our potential or underestimate our uniqueness. For better or for worse, we influence the earth more than any other species. More often than not our historical influence has been destructive. Our survival and the survival of countless species now hinges on making the changes in consciousness that will end this cycle. The *Starseed* material addresses *the habits of perception* that lie at the root of destructive human behavior, and the moment-to-moment awareness that can bring our influence wholly into the realm of the creative.

Except for the occasional accessing of individual memory records, the awareness I experienced while transcribing this series did not differentiate between I, you, and we. It was unable to comprehend our over-literal use of these terms, and seemed to regard language rooted in such rigid divisions of self as little better than primate babble.

Whatever success I have had expressing an awareness so antithetical to the implicit assumptions of language would not have been possible without the help of intermediary forms of intelligence. Over time I came to think of them as "informational beings." What they truly are is not so easily labeled. They described themselves as "condensations of consciousness" in a sea of universal being, temporal embodiments of eternal qualities, ever changing in form and content, yet consistent in the qualities they represent. Introducing themselves to me in the opening chapters of *The Starseed Transmissions*, they explained:

> The eternal Being we represent is able to focus attention on many things at once. We are that attention; we personify eternity's interest in time. Our individuality is defined by the unique nature of the space we enter, but our intelligence flows from a common source, and we re-

4

main unified with that source in essence and in spirit. Although your human individuality is more specified and relates to material levels of manifestation, it is fundamentally no different. The chief difference between us is that you have forgotten the singular being at your core, and we have not. We have come to remind you: Humans are not isolated entities. You are cells in a universal whole—intelligence cells, now dormant, but with mind/brain circuitry designed to reflect the thoughts of the creative spirit at the source of your life. As you release the fictitious beliefs that stand in your way, memory of these things will return. You will recognize eternity's thoughts as your own, and activate historically latent features of human design. You will see that creature and environment are inseparable, and that environment does not stop with an ecosystem but extends to include your star-system, and every galaxy resting in the universal sea.

This explanation left little doubt that I was in contact with a form of intelligence far in advance of my own. Its description of itself and the others of its kind did not quite fit the stereotypes typically associated with angels, but over time it became clear that this was indeed what they were. As their primary spheres of influence are not of this earth, it is equally valid to describe them as forms of extraterrestrial intelligence.

Those who have sensed the presence of such beings at various times in their lives will no doubt find this Starseed series more accessible than those who have not. My hope is that readers in this latter category will consider this material nonetheless, accepting its stated origin as a working hypothesis, and—in the manner of objective science—postpone conclusions until the subject is further explored.

The information contained in these four volumes helps us view ourselves and our world from perspectives few of us have previously considered. It paints a portrait of a human race now poised upon the threshold of a breakthrough for which evolution has long been preparing. Collectively, that breakthrough still lies ahead. Individually, it begins the moment we raise our eyes above history's illusory divisions. The distinctions between self and other we once took for impenetrable walls then reveal themselves as filmy veils allowing passage to and fro, and we see ourselves reflected in all that meets our gaze.

This blue and white sphere, floating so alive through space, is, after all, an intensely intimate place. Its rhythms pulse within our bodies, our gardens, our seasons. As our minds and hearts also attune, our vision clears; we see the world anew. Plants arrange molecules in patterns inspired by the light of our nearest star. Oceans rise to greet the moon. A biosphere plays with endless variations, mingling earth and sun, time and eternity, balancing matter and spirit in uncountable lives.

Evolution moves toward balance *in consciousness* as well as in form.

Nature is a vast network of interwoven intelligence. These Starseed volumes address our inborn, though historically latent, ability to access it. They represent an early translation of the thought circulating in a unified field of awareness that has long been bubbling up through the earth's biology. They provide a glimpse of our possible human future and hint at the choices that may lead us there. Above all, they celebrate the extraordinary nature of much we have taken for granted: the heavens in our wildflowers, and the universes in each grain of sand.

Introduction

Although communication with disincarnate beings is not widely practiced in the industrial nations, such communication has played a central role in virtually all recorded human development. Since earliest times people have sought input and guidance from those "on the other side." In ancient Greece and China the oracles were consulted with regularity; their advice was a common feature of daily life. From the Bering Strait to Tierra del Fuego, native American tribes honored and revered the spirits of unseen ones. Both the Old and New Testaments contain passages dictated by beings without physical form. Human history, recent and ancient, provides countless instances of poets, saints, mystics, even statesmen and scientists who have experienced significant encounters with metaphysical beings of one sort or another.

Today, with increasing regularity, "channeled" messages from spirit beings are filtering into the media of the new information society, appearing in new movies, books, articles, videos, television documentaries and interviews. And the explosion of interest in channeling shows no sign of abating in the near future.

While in general I see this new wave of metaphysical input as a positive development, discernment of the value or accuracy of any particular source remains vitally important. One should never allow outside input, physical or metaphysical, to replace one's own better judgment, or assume that such input has any more validity than one's own firsthand experience. *Genuine* spirit information helps to broaden and deepen one's experience of this miraculous and largely unexplored universe. It never asks that one's own experience be put aside in favor of another's belief or ideology.

My own ability to access spirit world information did not develop until toward the end of a seven-year period during which my family and I lived in close proximity to the earth—tending an orchard, milking a cow, growing and preserving most of our own food, and providing for as many of our needs as was practical. During these seven years, which we humorously refer to as our "media fast," we lived without television, radio, newspapers or magazines. The news we received came from the earth, from the gentle passage of her seasons, and from the people around us. We had no electricity or plumbing. We hauled our water from a nearby spring, used kerosene lamps, cut firewood to supply our heating and cooking needs, and in general turned toward the rich fabric of our remote rural environment to find a deeper understanding of ourselves and the natural world in which we lived.

Though such a period of outward simplicity is certainly no prerequisite for developing spirit world sensitivity, those seven years—from 1973 through 1979—were, for me, a great help in learning to distinguish among the many types of channeling that can occur. I found that in the de-emphasis of outer media, "inner media" would come more and more frequently to my attention. I was both intrigued and fascinated. I used my leisure to examine it.

In time I came to think of my human body as a sort of biological radio that could be adjusted to receive different informational frequencies. I found that there were certain frequencies where insight of a profound and practical nature existed: insight that could assist me in becoming a more loving and compassionate human being, insight that could greatly enrich the quality of my relationships with my family, with the earth and with the society around me. I also noted lower frequencies that did no more than crackle with the static of collective human thought. There, early on, I encountered entities who quickly demonstrated a fact worth emphasizing at this time when so many are becoming interested in channeling: *there is nothing inherent in being disincarnate that _in and of itself_ implies the presence of any greater intelligence!*

However, on the finer and more subtle frequencies, it was a different story.

There I encountered beings in another league entirely, beings whose natures can only be hinted at in our current human terminology: spatial entities of extensive intelligence, inhabiting the higher frequencies of the light/energy continuum, living, breathing, having their very being in the eternal presence of that immense reality that we call God. These creatures are worth getting to know.

In fact, I do not believe there has ever been a time when the intelligence received from them has been more urgently required by a human society.

As our present generation juggles the multiple challenges of economic instability, nuclear weapons, the AIDS epidemic, the deterioration of the environment and numerous crises both present and pending, the thoughts, ideas and perspectives of these eternal beings can assist us in viewing our world through eyes less encumbered. Communication with them can be an important catalyst in the evocation of a whole new kind of post-historical thinking: thinking that instead of perpetuating problems, reveals solutions.

It does not require a Sherlock Holmes to note that our problems are rooted in our methods of thinking. And our thinking itself is rooted in assumptions so archaic it is doubtful whether they have received any critical evaluation since the last Ice Age. The problems of an analogous acorn society would not, in the end, be resolved by improving acorn methods of thinking, but by the beautiful and terrible bursting asunder that unleashes the magnificent power of the implicit oak.

The lens through which we human beings view reality has been locked into myopic focus for so long, it has rusted in place. We tend to view the physical through the physical, forgetting the spiritual and the eternal. With eyes caught in a tangle of revered cobwebs, we rarely notice even a single tree. The forest, for most of us, has become but a legend.

Brushing the cobwebs from our eyes and stepping back a few paces into the clearing of a larger consciousness, may prove to be more than just a perceptual exercise. Our future, our children's future, the very quality of life on earth may now depend upon precisely such an expanded understanding of the world around us. Given our present cultural biases, the initial attainment of such an expanded view of things would be about as likely as pulling ourselves up with our own bootstraps *were it not for the fact that we do have some help.*

During this past decade, a growing number of people have had encounters with those beings of the spirit world that Western tradition rather simplistically refers to as "angels." I have noticed, as have others, that these beings are not all of one type, but rather they belong to numerous tribes or families, each group having its own purpose and distinct sphere of activity. Certain of the angelic tribes, for example, rarely interact with the earth or her inhabitants. Others have been intimately involved with this planet from her inception, being, in fact, the agents and overseers of her organic development.

Within this second category of angels long associated with biological life is a highly specialized circle of beings who are responsible for the education of humankind. While these entities—who have been known on rare occasions to take human form—clearly have long-term purposes of a more general nature, their immediate goal is the freeing of our race from what they have termed "the spell of matter."

During the winter of 1986–1987, I was involved in a particularly significant series of encounters with entities of this kind. With but a few exceptions, these encounters took the form of tape-recorded dialogue sessions that were later transcribed by my life-partner, Sherry. From those written transcripts I have selected key passages, which are now presented in this book.

In the chapters that follow, I share first person accounts of several different angelic entities, some of them encompassing rather extensive perceptual fields in the sea of eternal being, others coalescing, like solitary crystals, individuating in a particular moment of time to incarnate for a season, or a purpose, or a day. On occasion, these individual identities offer highly specific memories and perspectives, but all of these perspectives spring ultimately from the same source. Each of them views the whole from a different point along the circumference of the great Medicine Wheel of eternal being.

They speak of the unique role of this earth in galactic development, of the purpose of a human presence on this world, of the rebellion of the warrior tribes, of the role of the Americas. And they speak of their own angelic, behind-the-scenes engineering as they have worked for millennia beneath the surface of human events, patiently drawing our race toward the very times in which we are now living. Through their varying accounts of past human history, present moment insight and events yet to come, their perspectives combine to reveal a meaningful and exciting new understanding of our current global situation.

One of the most interesting features of my work with these particular entities has been their ability to evoke vivid recollections of past events. In my earlier encounters, as recorded in *The Starseed Transmissions*, such a strong emphasis was placed on the importance of my attention being in the present moment, that this apparent focus upon the past initially took me by surprise. For some reason, I had never thought of angels possessing the quality of memory.

And time travel?

I had assumed it belonged only to the realm of fiction.

Yet, when I first relaxed into the electric presence of the great being who narrates the early chapters of this book and began to blend with it in a shared field of luminous perception, I was struck with what first appeared as a stark incongruity: in the acceptance of the present moment nature of its timeless spirit, I found that I shared its understanding of the past! I could view events of long ago as if I were watching some extraordinarily lifelike hologram. Relaxation into the presence of this angelic awareness had brought me into contact with what appeared to be a sort of collective memory, a detailed living record of human knowledge and experience. For me this was a first. After some initial hesitation, due primarily to amazement, I decided to take the plunge. I let go and opened myself fully to the experience of blending with the angel.

I relaxed in deep, inner communion with its powerful field of awareness. Together we became one with the Being whose unfoldment has become this universe. After a time of resting in the presence of that immense love, we swam out, as it were, to experience certain of the dimensional manifestations of that love as they appeared on a particular afternoon of a time now past, but strangely present in the moment of my viewing.

Pictures began to take shape around me. At first they were blurred and indistinct, but gradually they came into focus.

Incredulous, I watched as a dark-skinned figure in flaming headdress walked slowly over the crest of a hill, a sea of prairie grass waving gently around him in the sunlight of late afternoon. I felt as if I knew him, even as I knew the blue-coated soldiers toward whom he was slowly walking, arms extended in a gesture of peace.

In my mind a voice spoke. "Love, like energy," it said, "cannot be destroyed. Wherever love has found its way into dimensional expression, it leaves emotional alterations in the texture of the universe, traces like multicolored etchings in the ethers. Wherever two people have loved, wherever a human community has shared in the expression of love, whether it was yesterday, or ten thousand years ago, that love is still present, it can still be experienced.

"Events can be reconstructed from the impressions of love. They can be re-experienced with a lucidity almost indistinguishable from their initial occurrence. We open for you this window into an earlier moment that you might more fully appreciate the long centuries of preparation that have provided the context for your modern world. We have been sent to illuminate your understanding of the present human condition and to take you on a journey through significant moments of angelic/human exchange that you might more fully sense the great momentum behind the power that is now breaking through into surface human events. When you understand the past as we do, you will understand the inevitability and imminence of the Creator's awakening in the human family. You will be in a position to more fully cooperate with the Great Spirit's purposes, for you will at last recognize them as your own."

So began these communications.

Like a song through my window they came: pure tones on the wind, currents of awareness from a less familiar part of the realm. I welcomed them. I had no struggle with this material, no strain. I made no effort to understand, to remember, to believe or to disbelieve. When the communications were over, what belonged with me remained. What did not was nowhere to be found.

The purpose of these printed thoughts is to connect those who read them with the Source from which they flow.

And the home to which they call.

KEN CAREY
Greenwood Forest
February, 1988

PART ONE

1

The Bird Tribes

I remember the day when I walked across the open prairie with my head held high and my feathers blowing in the wind. The soldiers saw only my silhouette against the sky. I walked slowly toward them, arms extended from my side, palms facing them in a gesture of peace. I watched the waves of love emanate forth from my hands, as powerful as the love I expressed before and after Golgotha.

The soldiers shot me dead.

I knew they would.

But their children have been brought up on my teachings, have loved my spirit and have understood enough of my creative principles to sail to the moon. Could I have taught them in another way, when their bullets flew and my feathers blew in the breeze that day? Could I have spoken more plainly than through the example of my deeds?

I have died a million deaths and lived as many lives to teach the warrior tribes what they would not learn in any other way. In the end, I am the victor, because the warrior tribes are changing, fundamentally, while I am rising again and again, leading them and their kind ever onward toward their destiny among the midnight stars.

I live everywhere, all over the earth. I have memories to draw upon wherever there were gentle people through whose lives I knew the land. If I try, I can remember their place names, their faces, the streets of their villages, their dances around autumn fires when the forest floor smelled of dry leaves and moonlight filtered shadows through the naked trees. But other things, I do not have to *try* to remember, because those things I can never forget. I *am* those things.

I am often the mountain lakes, because these were the last places my people lived before they flew, before they left their human forms and took to the airs of spirit or realms of nature to wait for cycles and changing seasons to bring their time to the world again.

I could show you where five hundred people lived on the shores of one such lake. Yet you might glimpse a human only occasionally. As you might see an otter, beaver or raccoon, so blended were their ways and so in harmony with earth and sun their living. But the time of which I speak was long before the recent European migrations, long even before civilizational influence touched Olmec or Mayan heart.

Our cultures were pacific then in this undiscovered world between troubled Asians and warring European tribes. Your records speak little of the Americas, because until recently our cultures here did not create history. Our ways were simple. Our troubles—until about twenty-five hundred years ago—were few.

Only a few of our consciousness have dressed in human form these twenty-five centuries now past. Yet when we did, you could not distinguish us from the others. We did not fight. When I put on my headdress and rode my horse across the prairies, I was teaching. I was not fighting. I taught with feather-shafted arrows and landscapes that cradled the sunlight in a thousand sacred meanings. I drew the cavalry to the prettiest valleys to drink from the streams most likely to give them the truth, so that their children would grow up in the hills near those streams and eat the corn and summer squash that would teach them the wisdom their tribe had forgotten.

So I do not mind the time my hands were spread to the side, the soldiers fired and a body died. Lying there beneath the open sky, with the high prairie grass waving around me as they rode off in dust and disarray, I drew those bullets into something deeper than a body of soil and stream. I drew them into my soul. And my spirit flew to their source. I understood then what kind of factories made those bullets. What kind of women and children worked in those factories. I understood how they felt about their families as they were pouring the lead. How they regarded their land. What they thought of their fathers. Their mothers. Their grandmothers. And tumbling and billowing, rolling with them in their jumbled tribal consciousness like the towering thunderheads that massed over the prairie, I, too, dreamed their dreams, found in them what was true, and made it my own.

By digesting their metal words, I learned further of the teachings that would speak to the warrior heart. I taught them of electronics, of radio waves whispering wonders on the wind, of metal wings and material things that would lead them down the slow but certain path to wisdom. Again, as time after time before, I strengthened the educational influence that I have been weaving in, around and among their societies, while I drew them gently, ever onward, to their destiny.

You thought that you could shoot the dwellers of the prairies and forests and they would somehow disappear like a troublesome dream. You did not realize that they are my own, just as you are, that they would reappear in your children and in your children's children and, as your own people, live again. For like you, the native Americans of *recent* centuries have also been of warrior descent, with lessons to learn not unlike your own.

But this was not always so.

Once, we expressed through an American people whose societies saw our creations flower and flow like living brush strokes across the river valleys, lakes and forested mountains of two continents.

Listen and I will tell you of my kind, of the few who have remained to teach and guide you throughout your twilight age. I will tell you of our patient work as century by century, we brought the fearful ones closer and closer to the time of their deliverance and the time of the earth's salvation.

I will tell you also of the gentle tribes from which we come, the tribes whom some now call angels, but who in simpler days were known by another name. For thousands of gentle circlings of this earth around her star, the peoples of the Americas, the highlanders of Asia, the natives of central Africa and earth people the world over have known us as *The Bird Tribes.*

Except for a few of us who remained to guide and influence the course of your histories, the better part of the Bird Tribes have not been incarnate in human form during historical times, though always we were there in nature and in the spirit worlds around you. Some twenty-five centuries ago most of us departed, to wait for the return of your intelligence and the slow turning of the sacred cycles that would favor your awakening.

We are returning now, appearing like the first stars at the predawn clearing of a storm. Into your consciousness we come, appearing subtly at first, a whisper among your dreams. But steadily our presence grows, soon to shine as distinctly as the sun of morning. We come now with a message that will prove vital to you in these last years of history's night.

We call to you, human ones, beings of light. Rise up from your tradition-bound past and become our creative partners in the continuing unfoldment of wonder. Rise from your cultures to remember your origins and your nature: light bodies, like stars, agreeing to stay in specified temperature ranges for certain periods of time, creating time and space, distances between you, painting spatial landscapes on the screen of time, drawing stardust into the dancing fields of your light. Today you express in tongues and tanks and desert irrigation projects, in urban parks and towers of glass and steel, in ideas that people have and in ways that people live. *Yet your past is no indication of your potential.* Compared to an artist full-bloom in the flower of her time, history's accomplishments are but those of a child groping blindly in a darkened room.

When you, incarnate man, know that behind you is the same brilliance, light and warmth that expresses so passionately in the nuclear release of stars; and when you, incarnate woman, know that behind you is the power of the creative, the power of the new, the balance, the synthesis, the power of truth, then you will know that there is no threat to your life, to your existence. You will know that there is nothing to fear, for the immense powers of mother and father aspects of God have joined together in the creation of a material universe designed for your support and flourishment.

For three and one-half billion years on this earth, we of the angelic realms have fine-tuned and sustained the precise planetary temperatures most conducive to your growth and development. You have nothing to fear as you move into this cycle of cooperation with us and with your Creator. You are in good hands; and well loved.

Until recently, you were not in a position to understand these things. For so long as you opened your heart to the frequencies of fear, you used your power to give credibility to fear's illusion. We will help you now tune to the frequencies that will heal your world and draw you toward your future in the stars. Listen and sense behind these words the frequencies of love, where an eternal Creator's designs are ever unfolding in time.

We are the Winged Ones of heaven, your reflections in perfect love, the missing dimension needed for your wholeness. We are the spiritual guardians of the earth. Welcome us into your consciousness. Blend with us. Remember. Know yourself as fully, truly human.

The true human can tune to the voice of the river and give it expression. The true human can tune to the voice of the wind and speak the words that the wind cannot speak without human tongue. The true human can blend with the essences of the forest, the spirits of the rain, the spirits of every creeping, crawling, living thing and can represent them fairly and evoke from them the best that they can be.

Each form of created life is energy manifesting in matter, ever-changing, ever-flowing into something always capable of more expression, more unfoldment, more revelation of spirit world potential. The true human is designed to aid the development of all life forms, drawing out their ever-expanding capacities to provide always fuller revelations of that which lies in the heart of God.

The true human gives voice to the essences of all created things, but the first and central voice of the true human is the voice of the Great Spirit, the voice that says:

> I appear in sea, in wind, in soil, in starlight. In the sun I appear. I appear in mountains and desert rain. I am the star and I am the stone. I am bird and fish, sky and sea. One and whole eternally, I come, differentiating, multiplying, refracting like a ray of light through ten thousand water-droplet prism pearls suspended in earth's highest sphere. Shining I come, to touch the surface of this matter world in a splashing, multicolored chorus of light men and light women, created to bring beauty, love, order and grace to this sacred dance of atomic form.

Stardust. Frozen starlight.

You call it matter. It is an art form with which we have worked for twenty billion years. We have sculpted it into star systems, galaxies and a universe of rainbow-spectrum worlds. We are the children of light. We have been given the task of creating dimensional reality. We draw order, structure and beauty out of the vibrating music of starlight. From the redwood forests to the microbes, from the gentlest feathered bird to the most substantial ocean whale, our spirits manifest the Creator's designs in all the biological life on this planet. Yet only human circuitry has the capacity to embody the full reality of who we are, of who that Creator is—of who, in truth, you are.

Come, human children, the truth has always been here. Your prophets have ever made it plain. Wherever you have let us land, wherever you have let our consciousness settle upon the waters of your awareness, the open waters of your human hearts, ever we have come to inhabit your shores.

We are the superior intelligence that you are looking for in the galaxies. We have been in your parks, in your treetops, in your gardens, in cages in your livingrooms. We are the Bird Tribes. And we are returning. We are here to teach you how to fly. Take us out of your cages of concept and archaic definition. Come, join us in the living treetops, where the winds of spirit blow wild and free.

We are the Bird Tribes. And we love the flowing, fluid, gentle waters of our Mother's eternal truths, as we love the stars that nourish and support these biological robes of dancing light. We will never be far from water or land, from light or sound, from the Father's fire or the Mother's gentle wisdom.

For we are the Bird Tribes, entering this earth to swim beneath her waters as fishes of the sea, to walk on her land as the mammals and the humans of this age and to fly through her heavens on the wings of the Love that created us and creates us anew in each moment. Because we love these oceans. We love these lands.

We love this planet with an eternal fire that takes all the myriad stars to reveal in the fullness of time, but only one star to say to one world, "I love you this much, that I would give you all that I am; my fire, even to your ages of ice; all that I am, until we grow the children, until we conceive the offspring who will be equally a child of starlight's fiery love and an ocean world's gentle truth."

Awaken, humankind. The teachers of love circle round the morning star. Spiraling down. Coming to rest. They land. At the edge of your history's shore.

Fluttering into consciousness. The Bird Tribes return.

2

Carved in Stone

Remembered in African creation myths, honored by Australian aborigines, recalled in the folk traditions of every earth people, we are the Bird Tribes, the Solar Clan, the Children of the Stars. We are the source of all nativity, the way that spirit creativity comes into form in this world and in all higher worlds. Individuated. Intelligent. In spirit-being form, we are the tangible influence of an all-powerful Creator interfacing with a material world.

Releasing our identities in the burst of joy that translates them into light, we flew outward from our source stars, riding waves of light across the sea of space until we reached this world. *Each of us came from a different star. Yet our being is the source of every star.* And in that Being we are one.

In the early days, our work was to catalyze biological activity in the oceans of this world and to oversee the growth and development of incarnate creatures. We did this both when clothed in physical form and when remaining luminous beings of pure light. Throughout all of the earth's early ages, when it was necessary to clothe ourselves in physical form, we did so through a conscious process of materialization, a process that allowed us to draw into our luminous fields, in a very few moments, the atoms and molecules that would dress us in matter.

The forms we assumed during those early ages were of human shape, though lighter and less substantial than the bodies you wear today. The drawback of materialization, however, was that it took much of our energy and attention to maintain the bodies in which we dressed. So, though we had the advantage of physical instruments through which to create, sustaining those instruments was both awkward and distracting. We wanted to devote more of our attention to our creative activities and less to sustaining the bodies through which we were creating.

The solution, of course, was to create physical bodies that were self-sufficient, physical bodies that would have physical identities to care for them—egos, designed to work intimately with us, both in the maintenance of our human projections and in our continuing work of creation upon the physical plane.

To the first *biological* humans, we communicated this message:

> We are your spirits. It is our presence here in this river valley that has called you up out of the soil. You are unfinished, incomplete; you are still not all that you can be. Some of what you now embody represents development in the right direction, but some of what you now think of as yourself is in actuality part of what one day will become afterbirth. Let us help you sort this out in

yourselves, so that by emphasizing the value of your bodies as dwelling places for the Great Spirit, your minds and hearts will remain open to communion with us. In time, that communion will flow into blending and both you and we will know ourselves as one upon the physical plane. In union, we will know completion.

One specific star of all the myriad stars you see twinkling in your night sky sends its spirit to earth to incarnate in your individual human form. Your ego is designed to help that spirit (who lacks your familiarity with the cooler realms) to remain creatively incarnate here.

Until you have bonded with the star consciousness, you remain unfinished. You do not yet know your own material or spiritual nature. You will in time. But if we are to assist you now in peacefully experiencing this coming age while your species multiplies and fills the earth, you must remember to honor the values of our spirits, just as we honor you and your own material values. In time these distinctions will dissolve, and we will know ourselves as one; but for now we must tread lightly, extending to one another patience and respect.

The egos of the first incarnate ones welcomed this message. For a time, they were filled with enthusiasm. They genuinely loved us and shared our wish to see the Great Spirit's plan succeed.

However, as the moons of these first seasons passed, willingness to play their parts creatively did not last among all the incarnate ones.

What happens when an actor who was asked to play a supporting role rebels and tries to play the lead role instead? With no knowledge of the part?

You were there, in the beginning. You might remember.

Certain tribes among the first humans found the input of their five physical channels of perception so sensationally rich and awesome, that as their early centuries on earth passed, they forgot that perception was also available to them on intuitive and imaginative frequencies. Such oversight was significant, for the imagination and the intuition were the inner communication devices designed to keep each angelic spirit in close and frequent communication with its human projection during the early stages of bonding.

As these first human tribes placed increasing emphasis on the physical, their behavior came to orient more and more around sensual priorities. As time passed, certain egos began to assume decision-making authority in realms of spirit where they as yet had no experience. They began to place such a high value on caring for their bodies, they lost sight of the reason for having bodies at all.

A visible yet disproportionately influential minority of your early ancestors became infatuated with sensual input. They behaved with less and less awareness of our consciousness or the informing designs that we brought from the Great Spirit. The bonding that was to occur between spirit and ego could not happen in such a state. These egos came to see our spirits as "foreign" elements. They forgot that spirit and ego are two manifestations of the same presence, that each human body is created by the presence of a spirit being on the material plane. They forgot that they were reflections in matter of our spirit presence.

In time, the body and its physical priorities became their sole reality.

When a certain tribe began to think of cooperation with us as working with the invisible spirits of nonphysical beings (that they could not see with their eyes, but whom they felt landing, like huge birds, in the branches of their central nervous systems), they reacted in fear.

Instead of seeing our lighthearted spirit values as a profound basis for co-creation with their own durable physical values, the egos of this tribe began to fear this healthy and quite compatible difference in value systems. They gave exaggerated emphasis to their legitimate function of caring for the physical body and began to see our spirits as lacking in sufficient respect for the physical plane. They rejected our spiritual values, seeing them as frivolous and irresponsible. They identified with a piece of the puzzle, with an aspect of their own wholeness. They did not realize that by doing this, they were in fact rejecting their true selves and choosing instead fictitious, mortal identities. They were focusing on the shadows and not upon the beings of light who cast them.

Seeing spirit as separate from themselves, the egos cut themselves off from the inner direction designed to guide the multiplication and flourishment of their species on earth. In these tribes, the incarnation of our angelic spirits was interrupted, the human creation unfinished. Without the egos' cooperation, the required spirit/ego bonding could not take place.

Human creation was still in process when the ego-ruled humans ran off into the jungle. We called after them:

> Whatever you do, do not cut off communication with us! Without the perceptual input of your invisible spirits, you will never survive as conscious beings in the jungle of subconscious matter. You do not have a view of the larger picture. You won't see the forest for the

trees. Come back! You cannot remain conscious motivated by fear. You will experience sorrow, suffering and death.

But the ego-humans ignored our warning and departed into worlds of illusion. Their fear stirred up emotional turbulence, which crackled like constant static across the frequencies that were intended to connect us. Further communication with them was impossible. We had to let them go. Their free will was an essential ingredient in the larger, terrestrial gestation that was taking place.

As millennia passed, fear-centered egos came to dominate other tribes of early humans. Warfare was initiated. Love-centered egos declined in number. Eventually, fear-motivation supplanted love-motivation as the dominant human deity. And it would tolerate no other god.

So it was that the egos took the center stage of human identity.

The implications proved greater than anyone had dreamed. Human societies initially woven together by the gentle currents of tribal affection began to break up and decline. Like a beast issuing ignorance, the savagery of the cave man era was released upon the earth. The spirit world blueprints, the Great Spirit's operating instructions for the human body's proper use were ignored, the earth's guidance system, short-circuited. Our communication systems, designed to connect us with our creation, were now inoperative. We were left with no choice but to guide the rebellious tribes externally and trust that eventually they would learn.

We longed to see our people running lightly through the forests and meadows again, creating songs that would bring joy to the earth and her creatures—not struggling in pain, building their huts and hovels along crowded river banks. We longed to see the warrior societies stop hurting one another and crying out desperately to images of a God now carved in stone. We longed for them to start seeing once again with their own eyes the spirit of the living God that is both the Creator Mother Earth, and the eternal Spirit Fire of the Sun. We longed to see them welcome us back to complete the design for the optimal development of our sacred world.

But your ancestors were difficult, oh so difficult back then to reach. For when human beings cut off their conscious connection with the Great Spirit, they enter the twilight realms of the subconscious where evolution, education and precise justice prevail. The lords of karma meticulously rule these twilight worlds, like well trained umpires policing every wrong and right, seeing that the books are balanced at the end of every age.

For the sake of saving as many of you as possible, closed systems—tribes, nations, even whole continents—were isolated to preserve the rights of the inhabitants to choose their own forms of education.

Tens of thousands of years passed, containing many epics, histories and adventures.

We of the angelic realms preserved the harmony in as many lands as we could. For thousands of years, we exercised a benevolent stewardship over the warrior races. Many traditions still recall our era of stewardship as "the age of the gods." But as your ancestors multiplied, gradually the lineage of the Bird Tribes receded. Gracefully, we gave space for your cycle of learning to unfold.

The few incarnate ones whose egos did still love and serve the spirits of God, quiet, peaceful people, took to the hills, the mountains, or the islands of the world when the warrior tribes became too troublesome in their native lands. In some of these tribes we were fully incarnate. In others, our guidance was sought, but we were seen as external beings, separate from themselves. The former we guided from within, the latter from without. With them, we retreated to the higher places, and, if needed, to higher places still. Ultimately most of the gentle ones flew to the heavens to wait, choosing not to incarnate.

But even during the times when we were not able to incarnate in all, we were still educating, teaching and, as you will see, influencing the course of human events.

3

Civilizations

After the rebellion, there remained still a few human tribes who refused to participate in the rejection of their stellar spirits. In the sand around our council fires, we would depict each of our faithful tribes as a small circle. Around these Keepers of the Light Circles, we would then draw one great circle to represent the Great Spirit who included us all. We knew ourselves as the Ongwhehonwhe, which means "the people true to reality."

Due to our attunement to the original instructions of the Creator, and due, as well, to the energy amplification that we experienced through agreement among ourselves, earthly power was still very much in our hands. But some things cannot be coerced, and love for the truth is one of them. The cooperation of the egos had to be voluntary or the human design would not work. Though we had the power to do so, eliminating the warrior tribes in some great battle and thereby driving them from the earth was not something we considered. Incarnate or disincarnate, the warrior tribes remained a part of us. Incarnate or disincarnate, they would continue to disrupt the emotional atmosphere of the collective consciousness that, with them, we shared.

The physical level was the level on which the warrior peoples had fallen. It was the only level on which they could be made whole. Incarnate lives were required to teach them the folly of fear-motivation. Nothing but earth experience could bring them to a time when a core of them would again turn their hearts to God. Their passionate spirits remained essential ingredients in our future global species. We had to accommodate them.

We gave them the lands surrounding the sea that is today called the Mediterranean. Elsewhere, we maintained our presence.

But the warrior tribes were aggressive. They multiplied rapidly. Our way was to teach through patience and example, not through warfare and confrontation. So as millennia passed and they took over more and more of what is today Asia, Africa and Europe, we gradually withdrew. In the end, there were only two islands where we were able to maintain a consistent presence into historical times.

Two islands. Each an entire continent.

Wild. Uncivilized. As refreshing to us as the abundant fresh water that flowed living through their heartlands and rained amply upon their lush vegetation. Over these two continents, we sustained a clear and protective vibrational field in which many tribes lived and prospered long after the rest of the human world had plunged into darkness and disorder. You of this present generation know little of what transpired in these lands. There was no history. No written record.

The people of the Bird Tribes came and went off the face of the Americas without a single name being carved in stone, with not one stone head being raised in arrogance to look out over the sea, with not a single human bone left to note them in the Mother's soil; for they ascended, these tribes—and left no remains.

The Americas. Continents without turbulence. Few records. Few ruined cities. Of all the earth's lands, we remained incarnate here the longest.

After the rebellion, a certain amount of warrior energy had penetrated throughout all the fields of thought and emotion that surround the earth, but in the Americas, our continental auras filtered out the most turbulent of these influences. On those rare occasions when excessive fear-motivation did find its way into expression among our people, we would hold a meeting of the council. If the people were a danger to the sacred balance, they were asked to change their ways. If they refused, they were asked to leave. Those who left went willingly. They would not migrate back. They knew the decision of the council was fair and just. They knew that they had chosen behavior that was vibrationally incompatible with our continental emphasis. And so they would depart. Either by boat, or by the great ice bridge to the north.

Those who left America to settle other parts of the world knew why they were being banished. If they preserved birch bark or buffalo hide records or maintained an oral tradition, their descendants knew that a time would come when they would be invited to return, when what was broken would be made whole again, when what was lost would be found, when what was forgotten would again be remembered.

Those of us who remained in America came and went peacefully, honoring our spirit selves, finding our balance with nature, living in harmony with the seasons and with each other. We were peace-loving tribes. We were not perfect, but we did not make a controversy out of every meal. We learned how to fill our needs in harmony with the earth and her creatures. We honored the land on which we lived. We saw the animals as our sisters and brothers.

The thrust of our creative interests was biological. We created fruit and vegetable foods: potatoes, corn, beans, peanuts, squash, melons, sweet potatoes, pumpkin, yams of many shapes and sizes. We created new life-forms to interpret sound and light in new ways, beautiful birds, intelligent creatures with whom we communed as fellow guests in the lodge of the Earth Mother. Our people knew the spirits of the plants and animals as equals and friends. There was still trial and error among us, fine-tuning here and there, but we had the opportunity then to fine-tune. Most of the human egos in the Americas remained true. Though the spirits of God were not fully incarnate in all, we were incarnate in many.

We could keep the fear disease out of our islands as long as human populations remained small, and collective human consciousness was not making itself felt on a planetary scale. However, the disease was a part of our collective field of being and the human populations attuned to fear grew rapidly in the millennia following the rebellion. Their vibrational influence became stronger with each passing age.

By about twenty-five hundred years ago, humans had multiplied sufficiently to signal the distinct beginning of a global human consciousness. And, not surprisingly, the fearful influences were felt more often in our lands. Our continental aura was weakening. The seeds of discontent began to sprout.

I was one of a small group who took on the task of monitoring the growing influence of the warrior energy in the thought fields that surrounded the earth. When I saw the seeds of fear-motivation begin to twist and warp the values of key Central and South American tribes, it verified what I had sensed. I knew the Europeans would soon be coming to American shores. I knew that we were approaching the cycle of purification and that soon, in less than two millennia, it would be a single human world. Our period of continental isolation was rapidly nearing an end.

So I made the choice. I would go to the heart of the warrior stronghold. I would speak to them myself. My teachings would likely be rejected. I knew that. But I would leave planted among them the images and concepts that would soften and modify their pending invasion. I would attempt to give them some conception of the overall global picture, to remind them of their own integrity, of why they were banished, and of the kingdom to come.

My challenge was to be in their world but not of it; for I knew too much, technically, to be in their world at all. I knew that it existed only in their imaginations, a creation entirely of their fears. So I did the only thing I could to truly reach them. I took my spirit and caused it to forget, that I might incarnate among them and grow up as a child in their culture, allowing my incarnate awareness to be tailored to their fiction, molded into a communication device that would speak their language.

I slumbered with them and filtered my awareness through their dreams. I put on the clothing of their culture, so that when I communicated, my teachings would not be regarded as foreign, but would be accepted as the indigenous wisdom of their own Judeo-Roman society.

I taught them, when I awoke, of the power of the living God, that it was ever available, that it was a power to which they could surrender all fear, a power that would infuse them, awaken them and make them whole. I showed them it was not an abstract, mystical, or etheric power, but a practical, biological reality, living within them: the power called Life.

I taught them that the life within and around them contained profound intelligence and design. I showed them how to contact this intelligence and how to allow life's designs—the kingdom of heaven—to emerge from within their own hearts.

I taught them of the spirit beings, the angels, who have been wielding the creative power of life on earth for three and a half billion years. I explained to them how blending with the Holy Spirit of these beings would transform and revolutionize their lives, but that they themselves had to invite spirit into their affairs; for without invitation, the Bird Tribes would not interfere with their freedom of will.

In my life, my teaching and my example, I made the point that human beings do not have to be controlled by their fears, that they could relax and surrender all their fears to the Fire that burns eternally in the sacred heart of God, that they could align themselves with eternal love, become one with the Creator, and live their lives in peace, in harmony with each other and with a benevolent universe.

There was something I cared about more than anything in the world. I longed to see the human beings incarnate in the warrior societies forgiving one another, loving one another and looking at things through the eyes of their spirits instead of exclusively remaining locked into the perceptions of their egos. I longed for them to come to an awareness of the magnificence around them, of their potential and forgotten joy. I called upon the highest good; and, because I was aligned with the Creator's intentions when my statement was cut short, my prayer triggered a response from both heaven and earth. My call broke through to the highest levels of the spirit world. Others saw the potential that I had perceived, and, because it was a matter close to the heart of the Great Spirit, *a context for healing the most violent of the warrior tribes was, in that moment, conceived.*

Any time a statement is cut short in the earthwalk of one incarnate and there is a longing to complete that statement, as there was in my case, it represents a bubble, a potential culture, like a yogurt culture or a yeast culture. This one happened to become a new sort of human culture, complete with the manifestation of every positive as well as every negative potentiality.

Christian civilization. Something that invited each one to complete the life the crucifixion interrupted, something that flew in the face of all odds, something that took the fear-centered darkness of the Judeo-Roman world order and split it down the middle.

There is an underside and an overside to what the unconscious centuries of Christian civilization have done. I know that well, for I and the other spirits of my Light Circle have monitored each of those centuries. Incarnate for days or moments in the lives of key figures here and there, inspiring the bards with songs of innocence and experience, we have guided those who were open to us: musicians, children, and leaders when we could.

Our historical, behind-the-scenes presence has been obscured from human vision because *historical people have no vision—they turn their perception over to others.* They refuse to give credibility to any experience not easily put into words. They believe that if their native language does not have terms for it, if people do not speak of it, it must not exist. In effect, they let others do their "seeing" for them. They give away their power. They become unwitting prisoners of a popular, language-centered fiction. They do not often notice even a single angel, much less an entire species of spiritual guardians, though we are ever-present behind the stage props of their preoccupations.

People paying more attention to what other people tell them than to their own perception is the beginning of civilization.

It is the beginning of a twisting, warping influence that ultimately produces a shift in human orientation: from the internal guidance of the Great Spirit to external dominance by others.

Civilization does not occur among healthy people. In healthy, non-historical societies, people are not overly concerned with other people's fantasies. They are certainly not manipulated through externally imposed descriptions of reality. In America, civilization was a step backwards. However, for the warrior tribes, civilization was a step in the right direction.

The domestication of animals and the growing of grains necessitated the development of values other than the values of the predatory hunt or annual raid on the neighboring village. So to the warrior races, we taught certain arts. The civilizations we initiated in Japan, India, China and elsewhere provided us with educational contexts in which we could introduce values that warrior societies would never develop outside such contexts.

Since their tightly woven linguistic preconceptions made it virtually impossible for us to communicate directly with the individual members of the warrior societies, we did the next best thing: we communicated, slowly but surely, through guiding the long-term development of their collective values. We had been working with this slow but relatively certain form of angelic/human communication for many long millennia before we sensed the time to introduce the more specific teachings that could only come from one of us who had actually incarnated among them.

The ideas introduced at the beginning of the Christian era, though initially rejected, nevertheless lingered on in Western consciousness. Like ivy working gently but firmly on stone walls, they have gradually taken hold and upset many of the traditional assumptions of the warrior state. By intensifying every creative as well as every destructive human influence, these ideas were directly responsible for the highly stressed social context that gave birth to the Industrial Revolution.

Although it has wreaked havoc with earth ecologies, the Industrial Revolution has been both valuable and necessary. When humankind is healed and we are able to work freely through our species, there is no ecological damage that will not, even before a few centuries, be entirely undone. It is easier to flush toxins from the earth's river valleys than to remove the toxic ideas that cause them from human minds. *In the order of healing, it is human consciousness that first must change.*

The most significant effect of the industrial era is that it has birthed the Information Age and has brought us into communication with your race on a scale far beyond what otherwise would have emerged.

The Industrial Revolution drew the inherent contradictions of the fearful state into high visibility. At a time when the intelligence level among human beings has risen to unprecedented heights, almost to the very threshold of awakening, global telecommunication systems now keep human attention focused upon the problems that result from fear-centered ways of thinking. It will keep your attention focused on these problems until you solve them through the only act that can: a total departure from the entire warrior paradigm. This departure from the old ways of history has begun.

Many are now aware of us.

Our consciousness flows daily through the word processors of a thousand creative thinkers. Our insights and perspectives are bubbling beneath the surface of dozens of new movies, hundreds of new books, and thousands of news articles and popular songs each year. Humankind is linked in a single socioeconomic community that now closely monitors all new and potentially relevant information.

Your present global civilization is rooted upon premises of separation from God as flawed as any civilization that came before, but it has dynamically accelerated our education of your race. It has increased human intelligence across the board. Hundreds of thousands of you are now about to understand something that two hundred years ago we could only get through in bits and fragments to a few, isolated individuals.

4

Survival, Cooperation, and New Beings

*J*ust as distinct, little beings work together to produce each complete human cell, *there are spirit beings designed to work in symbiotic cooperation with each human ego.*

There are many names for us. We have been called *angels, Bird Tribes, higher selves, hoksedas, spirits of the stars.* The reality of what we are is more than a name can convey. We are the multiple projections of the Eternal One, spirits designed to blend with present day humans. You now consider yourselves to be your egos, but egos are only one half of the human equation. The complete human is a spirit/ego partnership.

We are individualized aspects of one holy and eternal Being. We are your spirits as they exist apart from the spell of matter. *We do not evolve. We incarnate.* Our intelligence is pre-existent; our identities, highly focused frequencies of starlight. You need us to be whole, just as we need you to fully enter the physical plane. We seek to awaken in your minds and hearts, to incarnate in your bodies.

We are here to return your human circuitry to its Creator. Yet your human egos are the guardians of this circuitry, the stewards of your minds and hearts; and without your egos' cooperation, our bonding with you is impossible.

You ego-ruled humans who feud and fight among yourselves do not seem to realize the obvious: the very values that you employ to determine your social behavior, would, if employed by your body parts, effectively block any cooperative association of ribosomes, enzymes, mitochondria and other little life forms from providing you with even a single coherent cell, to say nothing of a healthy and integral human body.

Hundreds of thousands of little beings, all working voluntarily together, make a human body what it is. It is not a case of *survival of the fittest*, as your belief systems, based on short-term observation, proclaim. It is rather a case of *flourishment of the most cooperative*, as all long-term observation of the universe verifies. It is through cooperation with one another that diverse forms of life adapt and thrive. And it is only through the symbiotic cooperation of a multitude of simpler organisms that more complex organisms like your bodies are able to come into existence at all.

At critical stages in their development, life forms cooperate for their own advantage with other separate and distinct life forms. Over time their cooperation results in union. A new organism comes into being. Again and again this occurs in the formation of complex life forms.

This is analogous to what is about to happen again as your late 20th century human world reaches the optimal moment for materially oriented, ego-ruled human beings to be joined by their spirit world counterparts.

Your race is soon to experience widespread awakenings, or as some will see it, a massive descent of beings from the stars.

We are the dreams of the Great Spirit, the true dreams, the clear dreams, the pure dreams, never tainted by fear, never touched by any motivation other than the motivation of love; and yet, we are dreams nevertheless. Thoughts. Disincarnate beings drifting formless through a universe that has placed the highest value on form, living in the consciousness of a Creator whose desire is to take form in the very creature that will result from our bonding with you.

Our spiritual intelligence is the missing dimension, the rejected aspect, of your own wholeness. For thousands of years, you have been afraid of us. Many of you are now learning that you have nothing to lose and everything to gain by establishing contact with us once again. It is only your ego that makes you fear.

Your ego is here to look after your physical body, to make sure it gets enough to eat, to make sure it does not walk over the edge of a cliff or damage itself unknowingly. Your ego is the steward and potential master of all material-plane fears, an important and necessary component of your identity. However, your ego was never meant to provide you with your primary sense of self.

In a healthy state, the ego is a secondary component of identity.

In a healthy state, the Being behind all being, the self behind every self, the Great Spirit behind all of creation is experienced as your primary sense of self. Your ego does not have to be repressed or transcended for this to happen. It does not have to die. It simply has to assume an appropriate relationship with *the spirit that in truth you are*, the spirit that wants to incarnate and take up residence in your body/mind/heart system.

Your ego is by nature a reflection. It can either be a good, sharp, clear reflection, or it can try to be an independent reflection. Yet logic shows there is no such thing as an "independent reflection." Your ego may create such an illusion; but if you believe in it, you will be troubled and unfulfilled.

When your ego stops *trying to do everything all by itself*, and invites eternal spirit into your consciousness, your historical illusion evaporates like mist on a sunny morning. A polarity reversal takes place in the charge of your human envelope. The field of consciousness around you changes. Instead of your ego dominating your sense of identity and blocking your awareness of the Great Spirit, an eternal sense of self awakens within you. You know yourself as a projection of the Creator of all the stars in the sky. You know yourself as one of a family of god-beings, sharing God's Being. You remember. Everything is seen differently. The world is perceived anew through the eyes of a universal awareness. Your ego becomes your working partner, and you commence the conscious creation of a new human reality.

We bring to you an angelic awareness that historically has not often been incarnate in human form. We bring an eternal continuity of consciousness that henceforth you might know yourselves as we know ourselves. Together, in loving cooperation, we join to provide spirit and matter with the optimal balance.

We are not here to dominate your ego, but to secure its agreement, that between us we might work together and ultimately, as the years of this transition pass, merge into a single, biological, spiritual and psychological entity—an entity that shall in no way deny the needs and concerns of either ego or spirit, but that shall in every way honor the fundamental design and purpose of both.

We understand love and how it seeks to become objectified in a material universe. Your human egos understand the mechanics of the physical plane. Together we will form a single, creative dyad, an entry point into which the Eternal One's universal creativity will pour and from there, flow out to thoroughly transform these realms of matter.

We are coming now to help you make the shift from unconscious, creature worlds of biology, to conscious cocreative biology. We are here to make this great time of change as gentle as possible. Our purpose is to blend with your race, to bring you into harmony with the Creator and with the earth, to create a world that works for everyone, a world that allows for the optimal development of all creative potential. Many human beings are already consciously blending with us. Wherever human hearts are willing to honor the spirits of love, we descend, we incarnate. We complete the human creation.

During the age when fear was enthroned as the god at the source of human motivation, our incarnations only rarely occurred. Most of us—your symbiotic counterparts, the missing pieces that are needed to make you fully human—flew to the gentler, nonphysical realms of higher vibration, fluttering away as birds might fly, startled at the thrashing of some loud and ignorant creature, only now to return, to approach you again under more favorable conditions.

But even during the darkest times, those of one particular Light Circle—my own—would fly close. To influence. To inspire. To blend where we could. You have heard rumors of our presence. Teachers. Healers. Wise ones here and there.

Listen now, you of this present world, to the tale of one who once walked upon your plains.

5

White Buffalo Calf Woman

She came from the stars. To many tribes she came, though each knew her by a different name. You might see her now as two young Sioux hunters first saw her, walking barefoot, like a vision, across the low rolling hills of the prairie. They had climbed to the top of one such hill, looking for game, for some sign of movement on the broad rolling sea of grasses. Far away they saw a dot upon the horizon. They watched it carefully. By the time it disappeared behind the hill before them, they were almost certain it was human.

With bated breath they waited. At last, upon the crest of the hill, a young woman appeared wearing a beautiful white buckskin dress, decorated with dark porcupine quills. At her side she carried a skin pouch. An eagle feather, woven into her long, braided, black hair, caught the light of the early afternoon. Remarking upon her extraordinary beauty, the first of the warrior braves exclaimed how he would like to couple with her there in the sun-warmed prairie grass.

"Put aside such thoughts," spoke the other brave. "This is a sacred woman, a vision perhaps, certainly not one to be approached in that manner."

But to his surprise, the woman in white buckskin smiled at the lusty warrior and said to him, "Come to me. You shall have what you desire."

And so the second brave was left standing alone on the prairie, watching as his brother walked off, apparently enjoying the mysterious woman in the swirling cloud of dust that for a while hid them both from sight. When the dust had settled enough to see, there was the woman, bringing slowly together the stitches of her dress. At her feet, partially decomposed, lay a corpse, alive with worms, beetles and a cloud of hungry flies.

Then White Buffalo Calf Woman—who was the form in which the Great Spirit had come to teach the people of the plains—spoke to the young brave, who now remained alone and said, "A man who looks first to a woman's outer beauty will never know her beauty divine, for there is dust upon his eyes and he is as good as blind. But a man who sees in a woman the spirit of the Great One and who sees her beauty first in spirit and in truth, that man will know God in that woman; and should she choose to lie with him, he will share with her in enjoyment more fully than the former ever could. And all will be as it should.

"You, when you looked upon me, were not blind to my beauty, but your first thoughts were, 'Who is she, this beautiful woman? What is it that makes her countenance glow so in the afternoon sun? What thoughts are those that dance behind her eyes? From what land does she come? With what tidings?'

"And so, my young friend, have no fear. You, too, shall have what you desire.

"You and your friend symbolize two paths that the men of a tribe can take. If you seek first the sacred vision of the Great Spirit, you will see as the Creator sees, and in that seeing, you will find that what you need from the earth will come readily into your hands. But if you seek first to secure your earthly desires and forget the spirit, you will die inside.

"In olden days, most of the men took your path; but in this age many men are now going the path of your fallen brother. What you saw in the cloud was a speeded up lifetime; your brother lived many years in those moments while you saw only a swirl of dust. It was not so bad for him as you might imagine. He lived a life that many in this forgetful age would even say was a 'good' one. But he was ruled ever by his passions. In the end, his body turned to dust, for all his thoughts were dust. He had forgotten not only the Great Spirit, but his own spirit as well. He contributed nothing of meaning to me, to womankind, or to the people of the plains."

Then the young hunter asked the woman who she was.

With eyes black as the midnight pools between the stars, she looked steadily at him for a moment, as if her gaze alone gave obvious answer.

"I am the Spirit of Truth," she replied at last. "Your people know me as the Mother of the Old Ones, but as you can see for yourself I am not so old as all that. I am no older than any stalk of grass waving in the wind or any prairie flower. I am the great Mother who lives inside every mother, the girl who plays in every sister child. I am the face of the Great Spirit your people have forgotten. I have come to talk to the nations of the plains. Go before me to your village. Tell your chief to prepare a generous tepee that will accommodate all the people of your camp. I will be there shortly. I have some things to teach you, sacred things that your tribe has forgotten."

In awe and excitement, the young hunter ran back to his people and told his chief of all he had seen.

Now the chief of this young man's tribe was named Standing Hollow Horn, a name that no chief would have had in healthier days, for it well symbolized the hollowness and bravado that by this time had come to typify all too many men of the plains. Yet Standing Hollow Horn was not a thoughtless or an evil man. After hearing the young man's tale, he set to work along with others to build a great tepee of many skins in which the people of the camp could gather; for the weather that early in the year was often wild and stormy in the evenings, and Standing Hollow Horn would not have the people distracted from the teachings of one whom he suspected might well be the mother of stars.

When the people saw White Buffalo Calf Woman approaching them across the prairie, they were amazed. They were expecting one of greater years. Yet she appeared no older than a maiden, as graceful as the grasses that waved around her in the twilight. Her face glowed with a radiance that spoke of wild flowers, watercress and the finest of herbs.

As she walked through the center of the village, many a warrior had first thoughts like those of their fallen brother. But they had heard by now the story of his fate, and they practiced a discipline of thought that to some of them was new.

Barefoot, as she walked always upon the earth in her travels, White Buffalo Calf Woman entered the tepee of many skins, her white buckskin dress radiating the presence of her spirit. Without speaking a word, she walked slowly in a circle around the fire that burned in the center of the tepee. As each of her sun-browned feet touched the sand, all who watched felt her prayerful reverence for the earth.

Seven times, slowly and in silence, she walked around the fire.

Few could look into her eyes. Those who dared saw pools of perfect blackness. So large were her pupils that those who looked into them could see the mirror of their face and the reflected flickering of the council fire. And they felt that they were looking at themselves, seeing themselves as they truly were—not exaggerating their strengths, nor ignoring their failings, but as they were, naked and revealed. Those who could not with honesty look into the eyes of White Buffalo Calf Woman did not look at all.

Even before she opened her mouth to speak, her bearing and example had given the people of the Sioux the greatest teaching those present had ever received. When she spoke at last, her voice was like the song of waters singing upon the rocks, like the song of prairie birds calling above the meadows. It reminded those who heard it of the wind that whispered around their tepees at night and whistled through the branches of the mountain pines.

"Seven times," she said, "I have circled this fire in reverence and in silence. This fire symbolizes the love that burns forever in the heart of the Great Spirit. It is the fire that warms the heart of every buffalo, every buffalo calf, every prairie dog, every sage hen, every eagle and every human being. You people of the Sioux are like a single being. This tepee of many skins is your body. This fire that burns at your center is your love. The fire of your love is at times expressed sexually."

Here she paused and looked deeply into the eyes of those around her. "There is a creative as well as a destructive way for this to occur. Passion that roars out of control is like a wild grass fire that destroys everything in its path. But tempered with wisdom, that same passion can fuel whole generations; it can warm a thousand lodges through a hundred snowy winters and give its power to your children and to your children's young.

"Those, like the young man whose bones now lie beneath the prairie moonlight, who think first of the sexual expression of this fire and only second, if they think at all, of the spirit behind it, lock themselves into cycles of suffering and illusion—cycles that were unknown among our people a few short centuries ago, but which now are debilitating your nation, weakening your vitality and draining away your power.

"Creation does not take place where there is a scattering and dissipation of energies. Creation requires a gathering together and focusing of your power within a circle of commitment—like a seed, an egg, a womb or a marriage. If you would create and not destroy, you must remember always the Sacred Hoop. Consider wisely the ways in which you would use your power and then around those ways draw the sacred circle of commitment. In the warm atmosphere of that circle, the power of love builds and builds like a storm above the wet summer prairie until suddenly the circle can hold no more and explodes in the conception of the new.

"This fire is more powerful than any one of you," White Buffalo Calf Woman said, pointing as she did to the flames dancing in the tepee's center, "and the seven sacred circles I have walked around it represent the seven worlds that it has created. You here live in all seven of those worlds, but you are conscious only of one, the physical, the external. You have forgotten the inner worlds, the worlds of vision, the world from which I come and in which my kind lives. I am dressed like you people of the plains, but my tribe is not Sioux." She paused a moment and slowly bent forward to pull a burning branch from the flames.

"My tribe is the Fire Bird," she stated quietly.

"I am one of the Bird People, whose tribe once covered this Island of the Turtle. Do you remember the Winged Ones of heaven? The Fire-birds? The Thunder Tribes? How long has it been since you called to us in your councils?" She walked once slowly around the fire, holding up the burning branch and looking searchingly into their faces for an answer.

All eyes turned to the story tellers.

The story tellers were silent.

"Your people have forgotten," White Buffalo Calf Woman continued, "that which is more precious than water. You have forgotten your connection with the Great Spirit. I have come," she said, holding the burning branch above her, "with a fire from heaven to kindle again your memory of what has been, and to strengthen you for the times to come."

She then placed the branch back into the fire and took from her side the skin pouch she carried. Many only now noticed this pouch for the first time. They were amazed at the beautiful beadwork and porcupine quill designs on its surface.

"In this pouch," she said, "is a pipe. This pipe is sacred. I give it to you to help you remember the teachings that I bring. Always treat this pipe with respect. Carry it and the others you may make after its fashion only in bags of the finest skin, decorated by only the most reverent of hands."

Not yet opening the skin pouch that contained the pipe, but placing it reverently near the fire and occasionally gesturing toward it, White Buffalo Calf Woman explained the use of the pipe.

"Fill this pipe with a sacred tobacco grown especially for the purpose. Draw your first breath of smoke from this pipe as a breath of gratitude to the Great Spirit, from whose breath you were first given life. Use the smoke of this pipe to represent your thoughts, prayers, and aspirations. Send them upward with your exhalations to the Great One, Wakan Tanka, Grandfather of all. Each time you do this, pass the pipe slowly and reverently among those who may be gathered with you while each offers his first inhalation to the Great One above this world.

"Then with your second breath of the sacred tobacco, let your thoughts be of love and gratitude to your Mother, the earth. Give thanks for the grasses that clothe her breasts in prairies of flowing grain. Give thanks for the canopy of blue sky that she holds for you as a world in which to live. Give thanks for the storm clouds that bring rain to the prairies, filling creeks, water holes, springs and ponds. With reverence pass the pipe around the circle while each one takes a second breath of the sacred smoke and does the same.

"Let your third breath be for the four-footed and the feathered ones, for the buffalo and the prairie chickens, for the fishes in your rivers and for all the creatures of this good earth.

"And let your fourth breath be for the Ongwhehon-whe. Let your prayer be that your tribe will always remain among them, and that one day the people who remain true to reality will include all the nations of the world."

All this time she had not yet opened the pouch in which she held the pipe. Now, she slowly untied the leather thongs that bound it and lifting a corner, reached within to pull from the white skin pouch a pipe of red stone. The manner in which she raised this pipe for all to see spoke of such reverence that all within the great tepee grew still. Many in that moment found their hearts full. Tears glistened in many eyes.

"This sacred pipe," spoke White Buffalo Calf Woman, "and every breath of sacred smoke you breathe through its stem will help you remember that every breath you take is sacred. The bowl of the pipe is made of red stone. It is in the shape of a circle. It symbolizes the Sacred Hoop, the sacred circle of giving and receiving, of in-breathing and out-breathing, in which all living things come to life through the power of the Great One."

Asking for some tobacco, White Buffalo Calf Woman filled the pipe saying, "This tobacco I know was grown in your most fertile soil and given your most special care. It symbolizes the plant world, the moss upon the stones, the flowers, the herbs, the leaves of grass that cover the hillsides lest your mother lie naked in the sun. You, my people of the Sioux, are here to care for the earth. Your life is lit from that same fire that burns in the heart of the Great

Spirit, Wakan Tanka." So saying, she stuck a small twig into the fire until it burned brightly.

"Just as I light this individual twig from the great fire that burns in the center of this tepee, so each individual human being is a flame taken from the eternal fire of God's love."

Slowly she moved the flaming twig away from the central fire and held it up for all to see.

"This, your individual human life, like the single flame that burns upon this twig, is sufficient to light a great fire. As long as the love that burns within you is turned toward self-centered pursuits, it will remain tiny like this flame. Remaining tiny, it will bring you no joy. Eventually, in the swirling winds of spirit, it will be extinguished.

"But when you live in harmony with the Great Spirit, your flame of love is fanned by those same spirit winds. You are in love with the very purpose of life! You light the fire of love in all you meet. You know the purpose of your walk through this world and you know why the Great One gave you a life flame: not so that you could keep your tiny flame to yourself, loving what you need alone, but so that you could give it away, and with the fire of your love, bring consciousness to the earth."

So saying, she held the burning twig until the flame was just above the red bowl of the pipe. To the pipe's center she touched the fire, inhaling softly through the stem until the tobacco glowed brightly. The first wafts of the sacred smoke drifted through the room. It was as if those present were seeing a pipe lit for the first time.

"Just as the tobacco that burns within this pipe of earth represents the plant kingdom," continued White Buffalo

Calf Woman, "so this buffalo you see carved on the stone bowl of the pipe represents the four-legged creatures that share with you this sacred world, Etenoha.

"These twelve feathers hanging from the stem of the pipe have come from Wambli Galeshka, the spotted eagle. They are to remind you of the feathered races with whom you share the great circle of the sky; but they are also to remind you of your spirit selves, the Bird Tribes, the Winged Ones of heaven. As I now pass this pipe to you and you give thanks to the Great Spirit with your first breath of tobacco, let these feathers remind you of the spirit beings who come from the stars to brighten your human lives. Let these twelve sacred feathers draw your thoughts up and away from the gravity of petty and jealous passions. Let your thoughts fly, like these feathers have once flown on Wambli Galeshka, high above the world of the little self.

"Take this pipe. Give your thanks to the Great Spirit and pass the pipe to the others in our circle. Let your thoughts be lifted up to the Great One who comes now to stir your memories and to open the eyes of your story tellers.

"Every dawn that dawns red in your eastern sky, like the red bowl of this pipe, is the birth of a new and holy day. And just as the rising sun drives out the darkness, so the light that shines in the lives of all those who love, drives out the darkness of self-centeredness and dissolves the shadows that cause misfortune.

"Remember always to treat every creature as a sacred being: the people that live beyond the mountains, the winged ones of the air, the four-footed, the fishes that hide beneath the cool rocks in silver streams and lakes, all of these are your sisters and your brothers. All are sacred parts in the body of the Great Spirit. Each one is holy.

"The most difficult part of this teaching may be to extend this respect to the people of your neighboring tribes. Remember, like you, they are sacred people, given a specific work to do in the great Being of Wakan Tanka. Their work is not your own, their tasks differ from yours, but the purpose you serve is the same. The sun that shines upon you does not see you as being so very different. In peace you must live side by side with these who are a different shade of the color red.

"For a people are coming soon who do not share the color of your skin, but who are white like the snow that falls in the winter months. With them will also come those of black skin. And those of yellow skin. And those of colors in between.

"Just as the colors blend together in rainbows that arch across the prairie when the storm is passed, so you must tell the white races, the black races and the yellow ones when they come that though you are of red skin, you are first and foremost people of the Great Spirit. Through peaceful blending with your neighboring tribes, be an inspiration to the wandering peoples. Help lead all races into the harmony of the rainbow."

And then the pipe was passed, and she was silent until all those present had taken their first smoke. A second time the pipe was passed in reverence for the earth, a third for the four-footed and winged. A fourth time the pipe was passed in reverence for the many tribes of humankind, those in the distant past, those living now, and those yet to be. After all gathered in the great tepee had completed their fourth smoke, White Buffalo Calf Woman raised the pipe reverently before her for all to see.

"Carry your pipe with you always. Treat it as a sacred object. Honor all creatures and live your life in harmony with the Sacred Way of Balance of which every tree, every flower and every new day speaks. There will be many seasons when your heart will feel clear and pure as a mountain spring, and you will know the peace and joy of the Great Spirit. But should ever your steps falter, now or in the more troubled times to come, if ever you feel that you have stepped aside from the path of the Sacred Way, if ever your heart should feel heavy within you, do not waste time in regret. I will give you a ceremony," she said, filling the pipe once more and relighting it from the sacred fire, "a ceremony, which each of you can do in company with others, alone in your lodge, or out upon the prairie.

"Stop your activities. Find a rock upon which to sit. Asking for the Great Spirit's guidance as I have taught you, unwrap your pipe, and let its red bowl remind you of the sacred road, the way of life, the red pathway of the sun. After you have smoked your smokes to the Great Spirit, to the earth, to the animals, and to the people who are true to reality, after you have given your thanks to the four directions, then take a fifth smoke asking for the guidance of the great winged beings of the spirit world.

"Ask the particular winged being of the spirit world that is closest to you to help you see the wisest path to follow. Ask that spirit to help you make the clearest choice, to help you know the steps you are to take upon the path that your deepest knowing would have you travel. In time you will come to know that spirit being as your own true self.

For now, just rest in the still place where the deepest knowing makes its lodge. This will put you in touch with what you may have forgotten in the hurry of life. This will allow the fire that burns within you to speak to you in plain and unbroken terms.

"With this fifth smoke, the smoke that you offer to the invisible spirit that guides you, you will see that the spirit world is real, inhabited by wise and benevolent beings that watch over your trials and hardships, unable to offer you help or assistance until you ask them. With this smoke, ask the spirit beings that surround you to come into your life. Tell them you want to help them and the Great Spirit in their work and ask them how you can do this. By helping the Great Spirit in his work, you will help yourself far more than if you were merely concerned with your own affairs. Human beings are not fully happy or healthy until they serve the purpose for which God created them.

"Offer your sixth smoke to the six people whom you would most like to see especially blessed. A loved one whose spirit has flown from his body. A young man or woman who will soon be entering adulthood. The leader of a neighboring tribe whom you would like to see deepen in the ways of wisdom. Perhaps your own grandfathers, grandmothers or families. Each time you do this, choose the six people whom you would most like God to smile upon. For them, offer this smoke."

She then explained how the sixth smoke could be taken in six smaller smokes, one for each of the people involved. While the pipe was being slowly passed, she drew circles in the sand to represent the souls of each of these six people. Then around them all, she drew one great circle to represent the blessing.

When all present had completed their sixth smoke, White Buffalo Calf Woman turned slowly to look at each one in turn. All within the tepee grew very still. Even the outside breezes came to a rest and atop the poles, the tepee flaps hung motionless. The seventh smoke she explained must always be taken in silence; for it was offered to the Great Being from which every being was drawn. For that sacred mystery at the source of life, it was better, she said, to have no words.

After the seventh smoke was completed, such a unity was felt by those present in the great tepee of many skins, it seemed in that moment as if there was but one present. In perfect stillness, the silence spread like warm water, dissolving the last traces of disharmony from the heart of the Sioux.

For many moments White Buffalo Calf Woman was silent. When she spoke again, she compared her teachings to a tree, a tree that would flower in their understanding as they took these things to heart and applied them in their daily living. She told the Sioux how their individual human lives were like the individual leaves of a tree and how no tree has leaves so foolish as to fight among themselves.

"No people true to reality would be so blind," she said, "as to let the individuals within their tribe contend with one another. Remember, there is no cause that justifies violence against another human being, save only in the last resort to protect yourself or family.

"For a long time," she continued, "you will live beneath the sacred shade of this Tree of Understanding that I am planting in your consciousness this evening. And in the generations immediately following yours, the people of the Sioux will be united again in the Sacred Circle. Be true to that circle, my women of the buffalo, my braves, my people of the plains. Be true to these ways and hold steadfast in this truth. For a time will come like a dark storm from the east when the prairies shall be overrun with those who speak fast, perceive little, and wield much power.

"The sacred Tree of Understanding that you will carry within you during these next few generations will be cut down in that storm. The tree will seem to die. The Sacred Hoop will wither until it is all but forgotten. A few will keep the light of truth burning within their hearts, but the light will grow dim and become, even in those, but a tiny ember.

"But that ember will remain. Quietly, it will remain. All but forgotten, it will hold still a small glow in the hearts of the gentle ones. Even when a strange and hurried nation covers these prairies, even when your Mother Earth is bought and sold and stolen as if she were no more than a handful of beads, even when roaring ships of magic stone fly with people in them across the skies, that ember will hold still its tiny glow. And know this, my people: *a great fire can be ignited from a single, glowing ember!*

"For when the storm is over, that ember will ignite a dawn brighter than any dawn before. A new tree will grow, more glorious than this tree I leave with you now. With that new dawn, I will return. Under the shade of that new tree I will live with you. And with us will be gathered not only the tribes shaded red, but the white tribes from the North and the black tribes from the South and the yellow tribes from the East. In harmony, the four races will live beneath the boughs of the new tree. The age that we will see together will be the best that has ever been. All that has been broken will be made whole. The Sacred Hoop will be mended. The game will be plentiful and the spirits of all creatures will rejoice in the harmony of a new and perfect order. The Great Spirit, the very Thunderbird, will be active within the races, living, breathing, creating through the peoples of the earth. Peace will come then to the nations, for the original creators of life, the Winged Ones of heaven will return."

6

The Great Peacemaker

Our original vision saw the maintenance of integrity among American peoples. Had it succeeded, American/European exchange would have been strictly educational, with the native Americans teaching, healing and consciously leading the warrior tribes out of darkness. Had the tribes of America not departed from our ways, these past five hundred years of human history would have unfolded quite differently. It was not just the Europeans who shattered the peace; Europeans could not have transplanted their cultures on American soil had the peoples of the lakes remained true. And yet there is no judgment.

In the America of twenty-five centuries ago, one incarnate in human form with sensitive human heart was bound to feel the emotional turbulence of the times. Our hope was that the truth our people had lived with so long would serve them during this age of purification, that they would feel the pressure of the fearful collective in their hearts, but not allow it to control them, remaining a race of healers and teachers. In fact, it did not work that way.

In the people of South and Central America (where human predispositions had always been open, loving and relaxed) the growing pressure of collective human consciousness began to manifest as a strange new tendency toward fearful ways of interpreting the events of the world around them. This naturally brought their egos to the forefront. Self-centered behavior was increasingly justified until it spread and became focalized in distorted city states. Eventually, the people created what they feared. Exploitive, hierarchical societies appeared where for many thousands of years there had been decentralized, peaceful living in cooperation with the earth and her creatures. The people of the Americas had always had their challenges, but prior to this time they were external challenges, the challenges of rock and wind, heat and rain, challenging human ingenuity to meet them with successful designs. Until now the American tribes had not encountered the destructive challenges that arise from within when people are fearful and divided within themselves.

Once the people of the Bird Tribes lost touch with their eternal spirits, they were as responsible for what the Europeans brought as the invaders themselves. They were left with no choice but to blend with the warrior tribes, that in the end all might again become conscious and whole.

It was a dark hour, but not an hour without hope. The momentum of the larger cycles was still guiding all things inexorably toward eventual healing. Even after the fear-crystalization of Aztec, Incan, Mayan, Olmec, and Toltec peoples, there were still American tribes who remained true. Here and there, for a while, a few of us were incarnate and had influence. Quetzalcoatl. Deganawida. White Buffalo Calf Woman. Ulikron, the Golden One of the Sun. But as the centuries wore on and the troubles increased, one by one, we took to the air to fly into the spirit world, to withhold human incarnation until the great flowing tide of time should return again at the changing of the age. Bird Tribes do not stay where not invited. We do not impose. We blend only when our egos make us welcome. We flutter gently into human circuitry to offer what we are, to blend only where there is openness and trust.

We incarnated as long as we could in the Americas. But eventually the pressure of collective human consciousness, fearful as it was, created sufficient emotional turbulence that incarnation among the warrior societies became too painful for most lighthearted spirits of stars. By the time the first European settlers began meeting native Americans, they were meeting people whose true, native social order had been declining for over two thousand years.

By then, there were few of us incarnate as far south as the Great Lakes. Most of us had retreated inland to the mountains or far to the north. Toward the end we left even the north, without a trace, retreating disincarnate to the spirit world. We knew that the great, ever-rolling cycles would bring our time to the earth again. And most of us were content to wait.

I was one of the few who continued to put on material dress.

For a time I sought to draw the nations of the North American forests into a pre-existing reality that the Ongwhehonwhe had once experienced. I knew that my efforts could only partially succeed during this age when the dominant human vibrations were fearful, but I also knew that the mission I set out to accomplish could greatly reduce the trauma of the pending European invasion.

A confederacy of peaceful and cooperative tribes along the North Atlantic seaboard would provide a powerful center from which our energies could radiate during the troubled times that were approaching. Conceivably, our influence could spread from such a center, perhaps in time even reaching beyond the Great Lakes to touch the Mississippi and the People of the Plains.

In addition to the immediate value of the confederacy to its member nations, the principles that guided the confederate tribes would help to reduce the violence that would most likely follow in the wake of the massive European migrations, which were now but a few short centuries away. The confederacy would heighten the level of conscious exchange for all the people of North America. It would establish a precedent in human consciousness, a template, a pattern that would subtly affect human thought for generations to come. It was a goal worth attempting. Even partial success would be of substantial long-term value. I gathered my awareness and prepared for my descent.

I caused my spirit to glide across the forests of Ontario. From a tribe along the northern shore of the great inland sea, I felt the welcome of a family, strong and brave. I was born and raised in the Huron way. In a Huron village, I lived.

As he who is now called the Great Peacemaker, I traveled among the Iroquois speaking nations. I reminded them of the winged tribes available to help them and of the sacredness of all life. I taught them that warfare makes one as vicious as one's enemies and that to fight for any reason other than self-defense is to become what you detest.

It was to the fiercest of the northeastern tribes that I first brought my message, to the Ganienkehaga, the people of flint stone, or Mohawks as they are called today. Though these were among the distant, much later descendants of the original Bird Tribes, many generations removed, they nevertheless retained a deep, inner, rhythmic feel for the truths I presented. I knew that if the Mohawks agreed to the intertribal peace, it would be easier to win over the other Iroquois nations; for the Mohawks were famed for their bravery and abilities in war, and the other tribes were sure to take an interest in any significant news regarding them.

7

Among the
Ganienkehaga

"Who is this Huron to come here and preach to us?" the old warrior exclaimed. "Have these teachings been tried among his own people? Let him go back to his own kind. When his people have practiced these sayings and when peace has come to the tribes north of the inland sea, then let him return to us and perhaps we will listen. But now, these are empty words. Take them with you back to where you have come from, Huron. You are no longer welcome among us. Your words are not words of peace, but slippery words that cause confusion in the minds of my young men."

Without reply, I walked from the council fire.

I had been among the Mohawks now for some weeks. Even at first, they had only reluctantly tolerated my presence—an outspoken stranger from the north. Now, even what reluctant welcome I had was terminated. During the long, early spring evenings, I had been speaking to them of the needless tragedy of intertribal warfare. Around the central village fire, I had been presenting my vision of Mohawks, Oneidas, Onondagas, Cayugas and Senecas all living together with an agreement of peace between them.

However, in the course of these past weeks, my presence had brought many things to the surface, including a certain growing contention. Opinion regarding the wisdom of what I said was divided. The minority, who believed there to be truth in what I was saying, came into heated conflict with older warriors who greeted my every word with suspicion.

I might have still been granted more time to present my vision had there not been among the Mohawks at that time a bitter young visitor from a neighboring tribe, Ayawentha, an Onondagan, who for no apparent reason resented me from the first. Filled with hatred and quick to anger, he violently opposed every argument I ventured in support of a united confederacy. He pointed out to the Mohawks how my presence among them was itself a source of division and conflict. And that while I spoke of peace, I was in fact bringing disharmony to the tribe, creating disagreement and confusion. Ayawentha—or *Hiawatha* as he was more often called—maintained that this was both my purpose and my intention, that my words of peace were but a blind.

I slept alone that evening, away from the main camp.

The weather had warmed. The air hung heavy and expectant. I could smell the gathering rain. Late into the night I gazed into the cloudy blackness above, pondering my next step. Only once did I get a fleeting glimpse of stars. I awoke late having reached no conclusion. By the time I had hiked several miles to an area where it was known the deer abounded, the sky had filled with a confusion of dark clouds. The storm was near.

Midafternoon found me hunting along the edge of a deep gorge, lying in the branches of a large, dying oak that reached out over the edge of a seasonal creek bed some one hundred or more feet below. From my position in the over-hanging tree, I could see nearly a mile in either direction and thereby hoped to locate the current position of the White Tail herd whose trail had led me to the vicinity. This evening's discussion around the fire, I reasoned, would go better if I could supply the venison, for I still intended to pursue my mission. I would not let a few hard words deter the purpose of a lifetime.

I was surprised to hear others approaching through the forest behind me. This was a remote spot, yet I could tell by the rustling leaves that it was a large company. Two braves ran forward from the approaching party and gestured at me with quivering arrows to remain where I was. I found afterwards that Hiawatha had during the morning taken advantage of my absence to so rouse and enrage the princi-pal warriors of the Mohawk nation, including the chief and his closest counselors, that all of them had set out in a body to kill me, planning to thereby rid themselves of this man whom they now perceived as an enemy.

No sooner had the party assembled among the trees at the edge of the gorge when they began to berate me in the most violent fashion, accusing me of being a Huron agent sent to split the unity of their tribe.

I listened in patience at first. But then my spirit was roused. Ignoring the order of the two young men whose arrows were trained upon me, I stood in the central fork of the twisted and dying tree. I pointed out to them that though they claimed they wanted peace, they were so defensive and suspicious and unwilling to give those such as myself from neighboring tribes credit for having guileless motivation, they created enemies through their fear. I used myself as an example of one who had come in peace and friendship, but who now felt himself so mistrusted and falsely accused that were I any other, I would become, in fact, their enemy.

I told them how their own fear and suspicion had turned many of their neighboring tribes into enemies, when in reality these neighbors wanted peace as much as any Mohawk.

"Look," I said, pointing to the ground at the base of the tree, "I can see your hatred and your fear. Can you see it too, crackling through the air?" At that very moment the storm clouds, which for some time had been soberly massing above us, suddenly exploded. As a deafening roar of thunder shook the tree beneath me, a crackling fork of jagged lightning danced ominously upon the very ground at which I pointed.

"Look," I repeated in the pause that followed, "I will show you your hatred. It flows from the group of you standing there up the trunk of this misshapen tree and fragments, like these dying, twisted branches, into ten thousand twigs and brittle forks. The idea that you live with, that warfare among tribes is necessary, is as false and misshapen, as precariously rooted, as this dying tree that hangs suspended over a gorge.

"Your belief in warfare as a way of life not only divides you from other tribes, as the branches of this twisted tree divide from the trunk, but it also divides you from your families, from your own hearth fires, from all the things you love the most. And every day you go forth to fight, you know this is true. Do you increase your security through your belief in this monstrous lie? No! The idea itself divides you from each other, for it encourages values and forms of behavior that do not make for friendship among warriors, peace within villages, or happy songs around your sacred fires.

"This lie you believe is like a dying tree, a tree that offers no comfort, no fuel, no shade, a tree that saps the strength and twists the values of all those who become entangled in its many and devious branches. If you continue to live your lives among the roots of this deceitful Tree of War, you, like young Hiawatha there, will find the ravages of war entering into your very heart. Within your very soul will the battles rage as you are separated, as Hiawatha has been, from the inner peace that is the birthright of the Real Peoples.

"I am not a Mohawk. This you know. But neither am I the Huron you imagine now stands alone. I come from the tribe of the Real People. I am Ongwhehonwhe. I represent the unseen ones, the Winged Ones of heaven. They are with me now. The breath of their spirit fills my body, and my voice carries their power. Together we beseech you for love of all that is sacred and true, for love of your families, your children and your children's children, leave the deceitful Tree of War and take weapons against your fellow men only in self-defense. Move your village under the great boughs of the living Tree of Peace. Men of the Mohawk nation, if you care for honor, if you care for truth, if your children are to grow up healthy and whole, cut this tree of violence from your hearts."

"We will cut this tree of violence from our hearts all right!" Hiawatha spat in anger. "We will cut it here and now with you in it, to fall into the gorge where vulture, wolf and panther will find what little nourishment a carcass like yours has to give!"

There was a rumble of assent from the gathered warriors. An elderly chief raised his hand for silence and said, "Deganawida, you have been using this very tree upon which you stand to tell us, your betters, what we must do with our lives. We do not need you to direct us. We choose what we will keep in our lives. We choose also what we will cut away. I think this time young Hiawatha has spoken wisely. We will indeed cut this tree away," he said, his face darkening, "that you and your divisive words might fall with it into the gorge below where they will trouble our people no more."

Sharply, he commanded several braves to begin cutting down the tree. But with lightning movement before they could begin, I drew the tomahawk from my belt. Pointing with it to the darkened clouds above, I screamed a cry so savage and so terrible it seemed to some of the Mohawks that it was my command that caused the sky in that moment to open and the rain to begin to fall in angry torrents.

"No, warriors of the Mohawk nation," I shouted through the fury of the wind-driven rain, "it is I who have said the twisted Tree of War must be cut from your lives. It is I who will cut away this tree and fall with it into the gorge. My act will give you a teaching that my words have not. No man should ever be afraid to cut falsehood from his life, even if it is the very thing upon which he is standing. Once he recognizes it, he should not fear to let it fall away. For to remain standing upon a lie, once it is known to be false, is to forswear future peace and joy. And there is no value to living if such as these are your roots."

With savage fury I began to hack away at the trunk of the tree upon which I stood. The Mohawks were amazed. They watched in silence while the rain poured. With the passion of battle upon me, I worked my weapon ceaselessly, again and again in violent swings, each driving slash cutting deeper and deeper into the murky grain of the trunk.

Nearly an hour passed, punctuated only by the falling rain and the sound of my tomahawk against the wood.

I spoke no more until only a small section of the trunk remained. I knew that one or two more blows would sever the tree. I stood then for the first time since my fury was unleashed. This time the faces I saw wore a different expression. This they respected. This they understood.

Like all who are attuned to the Great Spirit, my timing was in perfect harmony with the natural world; for at the very moment I stood, the sun broke through the clouds and a rainbow appeared in an arch of brilliance across the valley behind me. I pointed to the circle of the sun. I pointed to the circle of the horizon. I spoke of the many complete circles of nature, the circle of the cycles of the moon, the circle of stars in the night sky, the circle of seasons, of fruit ripening on the vine, the circle of the honest and open eye.

"The truth is beautiful," I said. "The truth is like a circle. Yet the rainbow you see behind me is not a full circle, but a half circle, a half circle of five colored light tribes shining together in harmony. It appears today to remind the nations and the tribes of the Iroquois that we must complete the rainbow's circle in our lives, through our lives, through living in harmony with the earth and her creatures, through living in harmony with the laws of nature and through living in peace with the other nations around us. Each tribe is like one of the colors you see before you that together compose the rainbow. Watch for a moment. Do you see where the colors touch and intermingle? This is the trade between tribes that live in peace. Where people honor the ways of the Great Spirit, new and more beautiful colors come into being. Through their exchange, there is plenty for all. The rainbow has been given to us as a sign to remind us that we must become tribes living in harmony together on the earth. The rainbow is a half circle that we must complete in our hearts.

"Except for self-defense, there is no place for violence in a heart that honors the teaching of the Great Spirit. Remove such error from your lives, as I now remove this tree from its false roots."

And with that, I leapt into the air and landed with my full weight upon the overhanging mass. There was a tremendous crack as the twisted oak broke loose from the cliff and tumbled, with me clinging to its branches, end over end through empty space. The drop was over one hundred feet and the truth is, I had not before considered whether I should live or perish. The opportunity that had been presented to me to give that teaching was sufficient, I felt, to justify a lifetime. Whether I lived or died in that one body made no difference to me, for I knew that the people of the Mohawk would be more peaceful after what they had seen. I held tightly to the falling branches, feeling only gratitude for the perfect moment that held me in its heaven.

As it happened, the falling tree landed in the upper branches of several closely growing trees in the gorge below. Though I was unable to maintain my hold in the collision, my fall was nevertheless broken by a succession of branches. In reality, it seemed to me these branches were the hands of the Great Spirit. Catching me in his love. I reached the ground with a few scratches. A bruise or two. And with triumph in my heart.

As providence would have it, as I was making the return journey of about twelve miles, circling back up the canyon to the Mohawk encampment, a large buck crossed my path and ended up arriving with me on my shoulders as I entered the village just as dusk was falling.

The Mohawks were silent as I entered the camp. I threw the buck to the ground at the edge of the fire, making it clear that it was my gift to the people of the camp. I waited, standing in silence, only for a moment. When the same elderly chief who just that afternoon had approved the method of my death gestured to me to be seated, I knew the worst of the resistance was past. In the feast that followed, my surmise proved correct. The Mohawks were now open to hearing my message.

However, not one of the Flint People was touched so deeply as the Onondagan, Hiawatha; his transformation was like the transformation of darkness into light.

8

Adadarhoh

I remained all of the new moon among the Mohawk nation. They had now accepted me as one of their own. As I had hoped, the Flint People proved the first to agree to join a confederacy of tribes, provided I could secure the willingness of the other Iroquois nations. Curiously, Hiawatha was more interested than any other in my mission. Late into the evenings, long after the others had retired, he and I would remain to converse around the fire. As the weeks passed, we came to share a common vision.

When the time arrived for my departure, Hiawatha asked if he could accompany me. He made excuse for his former behavior by saying that he had fallen into a deep and black despair because his wife and children had been devoured by a monster-like human being called Adadarhoh, who was the terror of the whole region in which he roamed like a wild beast. According to Hiawatha, Adadarhoh was not the many-headed serpent that some believed, but he was just as vile: a wicked sorcerer of twisted body and mind who ate human flesh and hunted men as if they were animals.

Though reports on the nature and appearance of Adadarhoh did not often agree, and Hiawatha himself had never seen him, tales of Adadarhoh—or Tadodaho as some called him—had been told as far away as the Lake of the Michigan. Not a warrior from any of the nations that inhabited the northeastern forests would dare to journey into the dangerous vale where this twisted demon was believed to reside.

Since my work with the Mohawks was for the time complete and the region in which Adadarhoh roamed was in the general direction of the Oneidas, whom I next intended to visit, Hiawatha and I set out together on the journey, planning on our way to pay the murderer a call. The Mohawks thought it foolish to intentionally seek out Adadarhoh. Still, they respected us both and wished us well. They assumed that if we were not killed in the encounter, then we would surely kill Adadarhoh, for the presence of his evil was a menace to all the people thereabouts.

However, there is a spark of the Great Spirit in every creature that draws breath. Without mentioning the reputed snake-haired demon by name, I spoke with Hiawatha on the journey to Adadarhoh's vale of those times when strong and certain communication is able to accomplish more than the crudity of violence.

"When you simply destroy that which you cannot understand," I told him, "it will come back to you again and again and again, each new form worse than the previous.

But if once you can understand what motivates your enemy, you can often help him discover superior ways of accomplishing his deeper purposes."

"But you will never get such as him to listen to you!" exclaimed Hiawatha, knowing full well what I implied. "All he understands is the tomahawk. He is not a creature with whom you can speak logically. He understands only violence. That is the only language he knows, the language he speaks to every man, woman and child that crosses his path. I would do away with him, Deganawida, not for revenge—for you have helped me soothe those self-defeating fires—but for love. I would do away with Adadarhoh to free others from falling into the same fate that he wreaked so thoughtlessly upon my family."

And so we talked during our journey toward the vale in which this monster lived. And we made a plan. I proposed that I feign some sickness or affliction and try to secure the help of this Adadarhoh murderer, while Hiawatha watched us from a distance, his bow strung and ready to fire. My contention was that if this man had even the smallest spark of goodness in him, he would come to my aid and give me a basis for further communication. If not, if he turned on me, I would let Hiawatha have his way.

What happened proved better than either of us could have planned.

As I lay calling for help from behind a fallen tree, not far from the coals of Adadarhoh's last camp, the dreaded curse of the region came limping toward me with only the vile look of murder on his face. He approached cautiously at first, but when he saw me lying on the ground, pleading for a drink of water, he relaxed his manner. He looked carefully at the craftsmanship of my buckskin vest and leggings, which were, thanks to the Mohawks, newly made and of the very finest quality. Then, without a moment's hesitation, he raised his tomahawk for the kill.

But Hiawatha surpassed my expectations. For instead of shooting him, he leapt upon Adadarhoh from behind and knocked the weapon from his grizzled hand. In the same moment, I rolled, threw Adadarhoh off balance, then dove upon him, pinned him to the ground, and held him twisting, writhing and spitting curses. In less time than it takes to tell it, Hiawatha, a true marvel in action, quickly bound his hands and feet. Before he knew what had come upon him, Adadarhoh sat at our mercy, disarmed, propped against the base of a tree. In a taut bow, Hiawatha held an arrow trained upon his heart.

It was then for the first time that we truly saw this man whom all the tribes despised. His body was crooked and twisted, deformed from some hideous disease. Yet, his muscles, if somewhat grotesque, were nevertheless enormous. He exuded a strength in no way diminished but somehow increased by his deformity.

Though Hiawatha might have felt both emotions, there was evident not a trace of either anger or compassion as he spoke. "Adadarhoh," he said, speaking slowly and clearly so that the word quivered with a power not unlike the power of his taut bow. "Listen well, Adadarhoh, for I have you where your life is no longer your own."

And then, in grim detail, Hiawatha gave an account of the discovery of his murdered family. Throughout Hiawatha's narrative, Adadarhoh's countenance remained hard and set. He was no doubt certain at the end of this lecture he would feel that arrow penetrate his chest. But when Hiawatha finished recounting these events that had happened to his most loved ones, he suddenly threw his bow off into the bushes, disgusted with violence and all its spiteful forms. Savagely, holding the arrow in both his hands, he confronted the deformity tied to the base of the tree. With every muscle in his body taut, he held the arrow only inches away from the hateful face.

And then Hiawatha did a wonderful thing. He looked Adadarhoh in the eye and said, "I will forgive you and spare your life, but I ask that you do one thing for me in return. Listen to my friend, Deganawida. Consider what he has to say."

With that he snapped the arrow in two and lay its broken pieces upon the ground in front of the astonished Adadarhoh. Hiawatha turned to me then and asked me to speak. But my heart was too full. I shook my head and motioned for him to continue.

Then for the first time, from the lips of another, I heard my own teachings. Once again Hiawatha amazed me. He spoke beautifully, poetically, eloquently of the brotherhood of all creatures, four-footed, two-footed, winged. He spoke of the sacredness of human life, the joys of human friendship, the benefits that come to those who live together in peace and harmony and cooperation.

With eloquence far greater than I had ever achieved, and with a passion I recognized as springing from the same source as my own, Hiawatha spoke to that twisted ruin who had once been a man. And when he spoke of the benefits of a peaceful heart, I could feel peace flowing out from his own heart in great, soothing waves. I believe Hiawatha could have moved even the stones to sing when he told of the joy that fills the peaceful hearts of those who love the Great Spirit and live in accordance with nature's laws.

At first Adadarhoh listened in disbelief, looking for some kind of trap; but Hiawatha read his thoughts and seemed to address his questions and doubts even before they were spoken. As I watched, I realized that Hiawatha was a master at discerning the subtle meaning behind the slight twitches in Adadarhoh's face and the occasional movement of his eyes.

As Hiawatha told the story of my falling into the gorge, the countenance of Adadarhoh began to change. A dim comprehension grew behind his expression. For the first time, Adadarhoh spoke. His words were broken and halting. His face wore a puzzled look. He questioned Hiawatha regarding the location of the tree, for it had been a landmark thereabouts to those who hunted in the region, and apparently Adadarhoh had at least once stalked from its branches. When he began to grasp the fact that this tree had fallen with me in it into the gorge, that I had cut it off myself in order to communicate something important to the Mohawks, that I had survived the fall and that the entire Mohawk nation now regarded me as a sacred being, Adadarhoh looked in my direction with new respect.

In his broken manner of speaking, he questioned me about my tumble into the gorge and the details of my survival. He was amazed that anyone could have survived such a fall. His speech was not always easy to understand. He used a word that neither of us were familiar with, but that we gradually understood to be his term for justice. Apparently even in the system of his confused and twisted values, Adadarhoh had some dim comprehension of right and wrong. At last we realized that he was speaking of the justice of my surviving such a fall because of the strength and nobility of my intentions. He then gestured to himself and said plainly, and this time we understood, that in justice one with a heart as wretched as his own should not be permitted to live. He repented for the wickedness of his life and begged that Hiawatha should kill him then and there.

Hearing this request, I saw the flash of Hiawatha's long knife, as, without hesitation, he leapt at Adadarhoh and answered him not by killing him, but by swiftly cutting him free.

During the stunned silence that followed, as the three of us sat together in equal freedom, the expressions that passed over the bewildered face of the astonished Adadarhoh were awesome to behold.

Some minutes passed.

There was a humor in the air. A sense of brotherhood. A sense of all three of us at one time or another having experienced a certain darkness of the soul. Picking up Adadarhoh's tomahawk from the ground where it had fallen in the struggle that now seemed so long ago, Hiawatha spoke of his own recent transformation from anger and hatred to his present commitment to aiding me in my mission. As Hiawatha told his story, he made it seem as if such a transformation was quite a natural thing, as if it were no surprising thing, really, that one who so recently had been full of the violent passions of the warrior would now with equal passion devote himself to this great and noble cause of peace.

This was the first of many times that I heard Hiawatha speak in this way. It had its effect on Adadarhoh. In the months and years to come, I was to see it have its effect on others. There was a naturalness, not only in Hiawatha's person and in the form of his expression, but in the whole manner in which he conveyed his story.

Adadarhoh was slowly nodding his head in agreement as Hiawatha handed his tomahawk back to him saying, "We grant you your freedom, brother. Just assure us that you will not use this weapon against us or against any of the Iroquois peoples whom you will soon have reason to call your friends as I now call you, my friend."

Adadarhoh reached for the weapon. "Friend," he said meaningfully, looking first at Hiawatha and then at the tomahawk—as for a moment the two of them held it together. "Friend," he repeated. "I could never use this weapon on a . . . friend."

When Adadarhoh spoke the word *friend,* he spoke it slowly, as if it caused pain. Or stirred some ancient wound. Hiawatha withdrew his hand.

In the broken speech of Adadarhoh that followed came the story of his childhood deformity, the ridicule of the other children, his subsequent plunge into evil and banishment from his people. Adadarhoh spoke of these things, as though he were himself just remembering them for the first time after many long years. He ended his tale by gently lifting the tomahawk that Hiawatha had returned. Looking me and Hiawatha in the eyes, he told us again and with obvious sincerity that he could never use such a weapon against friends. By now we knew the truth: Hiawatha and I were the first human friends the man had ever known.

We stayed with Adadarhoh in his vale as spring warmed slowly into summer. I can recall the passage of no season before or since that I have more fully enjoyed. During the days, we would hunt and fish, at times together, at other times alone. In the evenings, around the fire, I would teach Hiawatha the songs that were the history of our people. As Hiawatha and I sang these songs together, songs of the Bird Tribes, songs born on the winds of eternal spirit, songs our ancestors had sung for tens of thousands of years before warrior values entered the hearts of our people,

ADADARHOH

Adadarhoh was visibly changed. He began to sit up more straightly and to think more clearly. He felt, as we did, the healing presence of the Winged Ones. As Hiawatha and I sang, often late into the night, we would sometimes see in the moonlight swirling, luminous currents flowing from winged figures that appeared around Adadarhoh while he slept. Sometimes we would hear the gentle strains of a fine and subtle singing that seemed to direct the glowing swirls with melodies not unlike our own. We knew that the Great Spirit was healing Adadarhoh's body. Words and example had already healed his understanding.

He did indeed, as it turned out, have an affinity for serpents, as one evening Hiawatha and I returned from a hunting excursion to discover his body virtually covered with writhing black snakes—no less than four of the friendly creatures entwined in his shoulder-length tangle of hair. It was an awesome sight to see them straighten their bodies in surprise at our approach, flickering their tongues and staring at us with tiny, intent eyes. Apparently, the snakes had provided the lonely soul with his only source of companionship during his years of exile.

When it was time for us to leave Adadarhoh to continue our journey to the summer encampment of the People of the Standing Stones, who by now had heard of our coming and were expecting us, Adadarhoh was a new man. Healed and strengthened, looking younger and stronger, he walked with us to the edge of the vale.

Before we took our leave, he held our hands together in his own and shook them saying, "I will go back to my own people, to the Onondaga who once drove me out. I will tell them what has occurred here and speak to them of what you have shown me. I will tell them of this great light,

93

of this wisdom you teach. I will tell them that two sacred beings will come among them in the not too distant future speaking of peace and bringing the songs and words of the winged power tribes that all revere. When you come, I assure you, they will welcome you and listen. You are my friends, you two. You are my first friends. I thank you."

We left Adadarhoh, hiking leisurely through a forest now sprouting the full growth of summer, speaking with one another as we walked of all that we had seen: of Adadarhoh, and especially of his people, the Onondaga, the fierce and stubborn People of the Hills. We were not to see Adadarhoh again for nearly three years. By the time we were at last to find ourselves among his people we would have already secured the agreement of two other tribes besides the Mohawk—the Oneidas and the Cayugas.

Though many tales could be told of what transpired during our journeys as we lived and worked among the People of the Standing Stones and the People of the Swamps, it was the conversion of Adadarhoh's people, the Onondagas, that was both our most difficult and most significant encounter among the five nations. Despite the growing strength of our reputation and the welcome that was extended to us due to Adadarhoh's testimony, the fiercely independent Onondaga proved to be the most headstrong of all the tribes, the most resistant to our teachings. They were ever polite in their listening; but being rugged people of the hills, they were slow to let mere words change the traditions of recent generations. Gradually, between Hiawatha's tale of my fall into the gorge and Adadarhoh's inspired (and much repeated) reference to "an older tradition that harkened back a hundred generations," we helped the Onondaga see their more recent traditions in perspective. Eventually they came over to our way of thinking and agreed to join the proposed league of nations.

During our work to bring the Onondaga into the confederacy, Adadarhoh proved essential. Without him it is doubtful the stubborn People of the Hills would have ever become the fourth nation to give their commitment to the Great Peace.

Apparently, Adadarhoh's central position throughout our negotiations and the high regard in which Hiawatha and myself evidently held him elevated him as well in the eyes of his own people. We were not surprised to hear some months later that he had been elected chief of the Onondaga. Eventually, he was elevated to a prominent position on the supreme council of the League of Iroquois Nations itself. This position he filled with honor and distinction throughout the remainder of his days. Though he was never noted for his eloquence, Adadarhoh's sincerity and genuineness of heart were unsurpassed. He possessed a keen and discerning insight into the motivations of men. More than once, his perception saved the confederacy from dangers others had not seen.

9

Tree of Peace

*T*he vision that first came to me one starlit evening some five winters ago was now rapidly becoming reality. During the past year the Seneca had become the fifth nation to agree to the peace. A gathering of the principal representatives of the Iroquois tribes had now been called to formally establish the league.

For months runners and braves had been traversing the countryside, inviting not only the chiefs and leaders, but all who cared to attend to meet at the headwaters of Onondaga, a sacred lake (in what is today upper New York State). They were told to gather around the largest tree that could be found four days before the full moon When the Geese Fly South. Virtually every inhabitant of the five nations who was fit to travel set out that fall for the appointed location. For weeks they had been coming, gathering, camping in anticipation of this historic and long-awaited occasion.

By the time Hiawatha and I arrived, there was no question as to the largest tree. It stood prominently in the center of a huge, natural basin at the edge of the lake. The surrounding area as far as the eye could see was filled with the encampments of the five nations. The basin itself, however, had been kept clear, for it and the open area around the base of the majestic pine had been designated as the site of the first supreme council of the new Iroquois Confederacy.

It was on a windy and sunny morning that Hiawatha and I arrived. In silence, the others gathered to greet us. With no words spoken, the chiefs and principal warriors of the Seneca, Cayuga, Onondaga, Oneida and Mohawk nations joined us in forming a circle around the base of the magnificent tree.

Surrounding our circle were thousands of men, women and children. They had come, not so much to hear the words that were spoken at this historic meeting—for voices did not carry much beyond the circle of the central council—but to feel the spirit of the occasion. Though there were many thousands gathered around us, so respectful were they and so silent that our initial prayers were broken only by the sound of woodpecker and chattering squirrel.

For three days we observed the *Wigwam of Silence*. From dawn until dusk, no words were spoken. On the fourth day, the *Wigwam of Oratory* began.

Standing upon a great moss-covered rock about thirty feet from the base of the towering pine, Hiawatha spoke with the eloquence that had served our purpose so well during the preceding years. He spoke of the Bird Tribes, the winged beings of light who come to earth, calling up these mortal bodies of flesh. He spoke of the law that rules throughout the natural world: the law that one receives in the course of a lifetime exactly what he or she has given, that one experiences oneself the same conditions that he or she creates for others.

Hiawatha spoke of the Great Peace that the whole of the Americas had experienced for hundreds of generations when this simple law was honored, respected and understood. All present could not hear his voice, but the leaders in the central circle followed closely every word that he offered; for by this time Hiawatha was regarded as the finest orator that perhaps had ever walked the Americas. Many of the chiefs were not only listening to what he said, but studying his manner, that they might bring something of his spirit to their own people as they in turn spoke of these truths in the years that would follow this momentous event.

Hiawatha spoke of a great lie that had become deeply rooted in the hearts and in the consciousness of all the world's people; a great lie, he called it, that had stolen the happiness and disrupted the harmony of the nations here gathered. Recognizing the opportunity for a symbolic teaching that could be witnessed by all present, Hiawatha referred to this great lie as I had once before: "The Tree of War," he called it. And throughout the multitude that surrounded us, the words were quickly whispered from one to another.

"The Tree of War," Hiawatha repeated, "has been as firmly rooted in the habits of our most recent generations as this pine is rooted in the earth. It may seem that a tree as large as this is too well established, too firmly implanted to ever be removed; and yet," he said, "this giant of the forest can be toppled by something as simple as this."

Here Hiawatha held up a single hand. And though every word was not conveyed to the multitude, the gist of

Hiawatha's meaning was soon grasped by those present. He paused until it was understood, and then he continued.

"Something as simple as a human hand, created by the Great Spirit with five fingers, each one separate, individual and independent, something this simple can topple this great tree when the five fingers work together as one. When our five separate nations are working together, we have great power. Let our five nations become the five-fingered hand of the Great Spirit. Let us eradicate the Tree of War, the habits of warfare, the ways of violence, not only from our nations, but from the nations we will influence in the generations to come. Human hands serving the purposes of the Great Spirit, given again to the purposes of peace, can uproot even this greatest of all trees in the forest."

We paused then, and I spoke quietly to the spirit of the tree, asking its understanding. Would it be willing, I asked, to release this particular tree-form and inspire another instead? Would it open itself to a death as glorious and meaningful as its life had been?

"Your life has been a good, long, healthy one," I said. "You have served the forest well. But the two-footed dwellers of this land need a teaching that only you can provide. They need this teaching that they, too, might serve the forest and its creatures as you have served. For the sacred cause of peace between the human nations, would you be willing, O Grandfather Pine, to open yourself to a death that will inspire all human races for generations to come, to be remembered in song and legend as long as there are people in this world?"

All assembled waited for the tree's response.

Slowly, gradually, we felt the emotional language of the tree speaking to our hearts. "Yes," the spirit of the tree replied, "I have long known that this summer was to be my last, that the time was near to release this form for another. Such a death will be an honor. I have been waiting for you, Deganawida. Proceed."

Then Hiawatha invited those present to bend down and using only their hands, to begin digging into the ground, into the forest soil beneath the needles and moss. And all joined and began digging a trench around the base of this tree that Hiawatha likened to the well-rooted Tree of War. He forbade the use of any tools or implements, but the enthusiasm of the chiefs was strong; and as they dug, their hands were joined by the hands of all those who were able to get near enough to be of any use. Soon a deep trench appeared around the base of the tree.

It was not long before the ancient branches rippled slightly as a light shiver ran through the tree from the ground upward, signifying the departure of the spirit. The first beginning creaks of instability were faintly audible around its base. Hiawatha held up his hand and called a stop to the digging. He invited me to step from among the circle. I walked through the trench that surrounded the tree and climbed until I stood beside the mighty trunk.

"Working together," I said, holding up my hand, fingers spread for all to see, "working together as the fingers of one hand, our five nations have great power, for behind us then is the power of the Great Spirit. When we align ourselves with the purposes of heaven and act in common accord, we have all of the natural forces working with us, supporting us, helping us.

"Though the truth may be simple, as simple as a human hand, and though it may be gentle, as gentle as a hand, aligned with the powers of heaven and the powers of earth, that single hand can stand up to the greatest lie, to the most entrenched habit or tradition, to the most stubborn and firmly rooted illusion; and that single hand can topple it as easily as my single hand now topples this tree."

During the time I had been speaking, many claimed they saw towering above the trees the great winged beings of light who bring the Creator's power to the earth. As I placed my hand upon the tree, many claim to have heard the rustling of their wings. But even those who did not see or hear these things felt the rush of wind that joined my human strength as the great tree, with a rumbling, crackling eruption, ripped itself from its roots in the earth and fell crashing to the forest floor.

The power and solemnity of the occasion were so marked, not one warrior's head turned. Not one human heart in fear skipped a beat.

Such a deep atmosphere of trust filled the assembled multitude, no one was injured in the fall of the tree. Some said they saw it caught by the hands of angels and gently lowered to the ground. There were also those who said in after-years that I had toppled the tree of my own accord. But so it is that stories grow. So it is that legends are made.

Neither Hiawatha nor I lost any time in making the fullest use of the symbolism. The fallen pine had left a cavity so large that a subterranean current of muddy water could be seen moving sluggishly in its depths. We invited all present to throw into the gaping cavity all weapons that had ever shed human blood or in any way been used in war. During the rest of that day as the sun slowly sank into the west, warriors followed the examples of their chiefs and soon the cavity was filled with a great assortment of weapons that would never again be used against human beings.

When the sun rose the next day, all hands joined together in throwing the rich forest soil over the cavity that now contained the discarded implements. Soon there was a great mound. When the weapons of war were buried beneath nearly six feet of earth, Hiawatha climbed to the top of the mound and spoke.

"Here beneath our feet is where all weapons of war belong, among the currents that flow in the nether regions, beneath the level of human knowing, beneath the level of our interest or concern. Let us bury these weapons of conflict, and release the hatred and mistrust that produced them. They do nothing for our people but cause suffering. They do not serve the needs of those who use them half so well as commerce, communication, brotherhood and cooperation. Let us bury these weapons from sight, never to be considered again. And in their place, let us plant a new tree, a tree of peace and understanding."

With a well-proportioned young white pine that I had carefully selected and prepared, I climbed to the top of the mound to join Hiawatha. As I held the tree, Hiawatha knelt and scooped out a place for it in the freshly turned soil. When the place was prepared, I knelt beside him. Together we planted the new tree, the Tree of the Great Peace, the tree that, in centuries to follow, became the most well-known tree east of the Mississippi, the Tree of the Great Long Leaves.

As the leaders of the five nations sat in a circle around the newly planted tree, Hiawatha spoke to them in a new way, in a new tone, with a spirit that he now drew from all our hearts, with a voice that was only made possible through the profound harmony and agreement of the thousands who had gathered with us.

"We are at the beginning of a new time," he said, "a time when the violent tendencies of human beings on islands far from the shores of this lake, far from this forest, far from any lands that we know, will grow and grow and become ever more restless.

"We are living in the beginning of the end times. It is good, because it is the beginning of the healing that will remove the dishonest tree of warfare from human hearts forever. But just as the darkest hour of the night is just before the dawn of the new day, so things will get worse before they get better.

"We are living in the beginning of times when the lies that turn hearts from the path of peace will whisper more vigorously in human thoughts than ever before. So it is fitting at the beginning of these times, when the powers of warrior ways will grow in the hearts of peoples far from these shores, right and fitting that we honor the precedent of peace established here long ago. We are living in the land where people from all parts of the world will one day come to be healed. It is appropriate that today we plant the tree of the peaceful way.

"This tree we have planted here will grow and flourish. In our own understanding, in our children's, and in our children's children's. For twelve more generations, this tree shall be strong; and during these generations much better ways will come to all the people who live beneath its teachings. The truths of this tree will flower, not only in Iroquois hearts, but in the hearts of tribes and nations far from these shores.

"But then a violent storm will come. A pale race will come from the East. They will be more numerous than all the game from here to the hunting ground of the Dakota.

And in this storm, which will last for five long centuries, the Tree of the Great Long Leaves will be utterly demolished. Not one twig or needle will remain to be found. Men will have even forgotten where it once stood. A time of sorrow far worse than any we have ever known will fill all our lands. The numbers of our people will become few. Our songs will be forgotten.

"Yet just when the grandmothers and the grandfathers begin to lose hope, just when the last storytellers have almost forgotten, suddenly it will be seen that in the very storm that uprooted The Tree of the Great Long Leaves, its cones were carried to the four directions, its seeds scattered by the far winds. In the hilly country it will have quietly taken deep root. But it will grow unnoticed at first.

"The white race from the North will bring with them a powerful black people from the South. And in time, from the direction of the setting sun, they will be joined by a yellow people. These three races will come from the three directions to meet us here on these very hunting grounds. Our children will mingle with their children and when the storm is passed, the ways of peace and cooperation will flower among the people of the world and spread to every land, and all the Great Spirit's children will live together in harmony.

"Someday," spoke Hiawatha, "these same thoughts that flower in our understanding today will reach out into the stars, beyond the farthest cloud and bounce back, echoing the teachings of the flowering tree. And all the nations of the world will hear these thoughts of peace and be amazed. They will know then the thoughts of God. The songs of the Winged Ones shall be sung again in the villages of the earth and the people of the four races shall after that live as one, even as the many leaves of this tree live as one. And there shall be no more war, nor sickness, nor untruth.

"I give this prophecy to tell to your children and to your children's children, to repeat around your teepee fires in the generations to come. When the great storm comes and it seems that the truth is for a time extinguished, hold fast, my brave ones. Remain true, my chiefs. Be steady, all you healers and children of the Great Spirit. For from this beginning that we make today will come eventually, when the storm of five centuries is passed, a new world where the peaceful ones will rule. And in that world, the barbarous custom of warfare shall have so far receded, even the storytellers will have forgotten it.

"Today we make a covenant with the peoples of our five nations. We make an agreement to honor the ways of love, the ways of justice, the ways of peace. We are not in truth a separate people. Each one of us here in this multitude is a single leaf on the Great Spirit tree. When we go within ourselves to touch the river of life that runs at the heart of our innermost being, each one of us touches the same life that flows within our sisters and brothers, even as the same sap flows through all the leaves of a tree.

"Your life is God's life. The Great Spirit is not just outside you but within you as well. Listen to the Great Spirit. In your innermost thoughts, you will hear teachings superior to any that I or Deganawida will ever share with the words of our mouths."

Hiawatha paused and bent his head while those present considered this truth. After a moment, he looked up and continued, speaking louder now so that his voice carried even to the distant rim of the basin.

"Just as you do not see the leaves on any healthy tree in the forest contending among themselves or fighting with one another, so neither is there any need for people to contend or fight or in anything but friendliness compete. When the Great Spirit is known within, in communities where truth and honesty are honored, people live in harmony with one another. They experience their own life and the life of the Creator as one. They know that our people are like the leaves and our tribes like the branches of a wondrous tree. They know that the life of the two-footed and four-footed and winged ones, the life of the plants and rivers and seas, the life even of the very sun, moon and stars grows also from that tree. All come from one single trunk of Being—the eternal being that we know as God. In this way, the Great Spirit lives within all things, within every plant and animal, every tree, every one of us here present."

As Hiawatha spoke these words, he moved about the mound, making descriptive gestures, drawing pictures in the air, pointing at times to the newly planted tree, evoking all the passion of his former warrior spirit. He channeled an enormous energy into conveying to those gathered not just an idea or a concept, but as much as he was able, an actual experience of the awareness that he himself knew in those moments. One could sense the chiefs as they felt it, understood it and experienced it each in turn.

From time to time, one of them would stand and, with a spirit nearly as beautiful as Hiawatha's, share a word or two of the great truth as it pertained to him and his people. But the chiefs kept their comments brief, for I think, like myself, they were truly awed by the Great Spirit whom Hiawatha in these moments had obviously become. Some say that as

Hiawatha spoke they saw a great winged form extending high above him into the trees. I saw only Hiawatha, but I saw him as I have known him in the Andes, in the Himalayas, and once upon the shores of the Sea of Galilee. There was no presence overlighting Hiawatha. The great eternal Spirit that is the reality of every human being had simply found a welcome home in his heart.

10

We Establish
the Confederacy

What happened next was remarkable in the midst of such a gathering of hunters and warriors. Some later said it was a miracle. However, to those of us who witnessed it firsthand, it did not seem unnatural. In fact, so natural did it seem and so much in the flow of Hiawatha's speech that our distraction was only momentary.

Hiawatha was speaking of the Great Spirit's life flowing not only through human beings but also through the winged and the four-footed. Even as he named them, a large snowy owl landed ghostlike in the lower branches of a tree immediately behind us and two does, followed by a less adventuresome buck, came ambling among the seated people—at times cautiously stepping over those who were in their way. Hiawatha pointed to the deer and to the owl.

"To peace with them also do we commit our living," he said, "taking life only when it is necessary to feed our people, giving thanks to the spirit of the animal whose form we eat, giving thanks to the Great One who appears as all these forms, within all these forms, in every tree, in every flower, in every herb-bearing seed, in every animal, in every stream, in every living, being thing.

"The Great One is here. With us, today, in these deliberations. As that Great One has both inspired and witnessed this ceremony, may he also inspire all our activities henceforth—everything we do, everything we say—as he has witnessed them from the beginning. May our lives and the lives of our respective tribes be dedicated to the spreading of this peace to the distant and far-flung peoples of the forests and plains. This is the finest way to honor our Mother, the earth. We are the people who remain true. Let it always be so."

Almost imperceptibly as Hiawatha spoke, twilight grew in the forest around us. One by one, lights appeared. A host of tiny fires gradually filled the basin of the new tree and danced across the nearby hills as people gathered near them for warmth. The full moon had not yet risen, but above us, as if reflecting our many little fires, stars one by one began appearing across the sky. A larger fire was built in front of the mound of the newly planted tree. Upon this mound, next to The Tree of the Great Long Leaves, Hiawatha stood.

"As the smoke from many family fires rises tonight and spreads above the forest, let us remember," he reminded the Iroquois, "that this smoke comes from many fires, and no one fire is better than another. Though one fire may burn brighter and another more faintly, who is to say which is wiser? You know as well as I that there are times and seasons for both. Remember always that the truth springs from many hearts and takes many outer forms, no two ever the same. One in the Great Spirit, we shall have no one leader, nor one ruling family, nor a class of ruling shamans, for such traditions are warrior traditions and not the traditions of the Ongwhehonwhe. Let such customs be buried with our weapons. Let them lie forgotten beside the tree that is no more.

"Under the Tree of the Great Long Leaves, the new tree flowering in our hearts, there is but one leader, but one chief—Wakan Tanka—the Creator, the Eternal One, who shines in the light of every star and burns in the fire of the sun. Know, my people, *that One lives in you!* Do not diminish the sunlight by giving your power away to another.

"From this day forward let all Iroquois-speaking people recognize that hierarchy makes sterile soil in which the Tree of Peace withers and soon dies. Wherever you have the entrenched leadership of one man or a small ruling elite, you have a society structured according to the ways of violence. Such a society is camped under the very branches of the Tree of War. Violence is the root of hierarchical society. Such societies will never know peace. Their people will always be restless, at war—if not with others, then within themselves.

"There is no hierarchy in God's sight. Such foolishness is unknown among the healthy tribes.

"Let no man be a leader for a lifetime; for I say to you, if any man has been a leader too long, he has clearly failed. The leader that wields the power and the magic of the Great One is leader only for a particular purpose, a particular moment, leader only for a season or a project or a journey.

"Not one of the geese I hear passing above is so foolish as to try to fly forever before the flock; his wings would quickly tire and he would surely fall. A spirit of such disorder would have nowhere to look for followers—except perhaps in the nether regions here!" said Hiawatha stamping his foot on the mound, "where these weapons and their ways are better forgotten.

"Let none of you think of yourselves as pupils, disciples, or students of either myself or Deganawida, for one who thinks of himself as a student never becomes greater than his master. And one who thinks of herself as a follower shall never excel her leader. Every one of you is capable of accomplishing all that we have accomplished here—and considerably more. What we have brought about during these last five years is but a small thing compared to what you will accomplish in the times to come. Look not to us to guide your feet on the path of peace, but look within yourselves and know the Great Spirit that is the source of our path, the source of our words and the origin of the deepest peace. Let his spirit guide you. You will never go astray.

"Though it may be fitting for a time, for a season, that the two of us are an inspiration to you, understand that in another season you will be inspiration to others and they in turn to others, and others still. And as the smoke from our many individual fires rises to heaven tonight and becomes one smoke flowing out from the Tree of the Great Long Leaves that we have planted here, may these thoughts that lie at the roots of peace spread out to all the nations and peoples of the world. Let the roots of this tree grow and spread out beneath this forest floor. I see them," Hiawatha said, pointing to the ground in front of him, "growing even now—Great White Roots of Peace."

Many saw in that moment the tree grow a hand's breadth or more. Beneath the firelight, others saw with an inner eye, four great white roots growing deep into the forest floor, spreading underground to the four directions, reaching into the subterranean currents that rule human passions. To many in those moments, the Great White Roots of Peace became tangible and real.

Above us our thoughts and ideas spread like smoke into the stars.

As Hiawatha sat down to rest at the foot of the newly planted tree, I stood. Behind us a majestic, orange moon was slowly rising, half revealed now above the naked trees. I spoke more slowly. I was conscious that each word I said was being repeated quietly, whispered from ear to ear. I could hear the repetitions almost like an echo or a gentle breeze rustling through the forest. I asked that one arrow be given to Hiawatha from each of the five assembled nations. All present could see the arrows, one by one, brought to the top of the mound and ceremoniously placed in Hiawatha's outstretched hand.

"Today," I said, pausing for the echo to take it to the farthest trees, "we have combined our individual power into one great power, the power of this confederacy. We symbolize this confederacy by joining together into a single bundle, these five arrows, one from each of our nations. At the base of the Tree of Peace, Hiawatha is binding the arrows with deer sinew, which is strong and durable. So shall our commitment to the Great Peace be strong, binding and durable. So shall these truths that we have celebrated on this occasion bind every one of us here into a single being that we might henceforth act together in unison, for the benefit of all.

"If but one arrow is removed from this bundle, it loses much of its strength. If two arrows are taken away, the power of the confederacy would be greatly weakened. If three arrows are removed, anyone could take hold of the bundle and snap it easily in two. Alone these arrows have little strength; even a child could break them. But bound tightly together in a single bundle, they are strong enough to withstand whatever might come against them.

"Let the Five Nation League face the future together. Let us think with one mind the thoughts of our Creator and always seek to peacefully resolve difficulties both within our League and without. Let us feel as one, with one heart, the compassion of the Great Spirit and strive always to avoid bloodshed. Should any one of us face evil from without, let us all turn and face that evil together, acting in unison with the strength of a single body. From this we shall derive great power, for from this day forward, we shall stand or fall, united as a single nation. We will leave this bundle of arrows beneath the Tree of the Great Long Leaves to remind us always of our union."

Stepping down from the mound, I invited all those who could to stand and form with me and Hiawatha a circle around the Tree of Peace.

"By holding firmly to one another's shoulders," I said, "we form a circle so strong that none could break through it. Let us live our lives in this circle of commitment to the Great Peace. Let there always be friendship, honesty and brotherhood between us. Let our children and our grandchildren grow up and live their lives in this circle of peace and security. Let nothing break or shake our commitment to the principles we have recognized through this sacred union.

"Always remember that each one of us in this circle is of equal standing and of equal power. With friendship and communication, with patience and understanding, there is nothing that we cannot in time agree upon. Do not let the white panther of discord come leaping and snarling into our agreement. For those who are quarrelsome and who repeatedly stir up trouble will be driven from the league like dead leaves driven from a tree in windstorm. And those who are expelled from the Five Nation League will not be invited to return.

"But my brothers," I said, "my sisters of the forest, let us hope we will have no need for such measures. For the Great Spirit has provided us with all we need to recover our senses should ever we fall into the confusion of blinded passion. Should any one of you ever feel angry or confused, resentful or afraid, climb to the top of the highest tree that you can find and look toward the Maker of this world for your relief. Call to the spirit tribes and wait for the answer to come.

"Let the thickness of your skin be as thick as the bark of a great tree, so that the barbs and taunts of others do not alter your balance or affect your center. Do not disgrace yourself by being known as one who is quick to anger, but let patience, tolerance and thoughtful wisdom characterize your life. Still, do not think badly of yourself if you do become angry or confused. Maybe it is only Gaha's way of telling you it has been too long since you climbed a tree! These passions are something that we all must face in the course of a walk upon the earth. But remember, we face nothing alone anymore, for this circle has joined us forever.

"You shall have no trouble that is not also my trouble nor any joy that is not also my own. I say this from my heart—and there are a thousand others with us here tonight who in their hearts say the same thing. If a thousand say this to you tonight, how many more will say it to you in the years to come? Remember that help is always at hand.

"You have friends in the higher worlds who will come whenever you call, and you have friends in this world that will do the same. From the highest tree, call out to the truth of your spirit and know that always, always, friends will come."

Then, amidst a rousing chorus of agreement, the circle disbanded and I resumed my place beside Hiawatha on the top of the mound.

"If any man or any nation," I continued, "shall desire peace within his heart, let that man or let that nation trace the Great White Roots of Peace to their single source in the being of the one great eternal Spirit whose presence lives within each beating heart. Let them take shelter beneath the long leaves of the flowering Tree of Peace. Let them live their lives in the shelter of these truths. For those who abide by these ways and principles will be upheld and cared for, their families will prosper and become a blessing to their tribes, to the four-footed, to the winged, and to the people of every nation they encounter.

"Let it be known around your council fires that leadership is necessary to tribal undertakings, for it draws together the power and the unity of the nation. But henceforth let your leadership be determined with the cycles of the seasons. Let the turnover among your chiefs and principals be as the turnover of fall leaves into the new leaves of spring. Since it is the custom of the Iroquois nations to have warriors and braves as their chiefs and leaders, and since this custom will not change until after the great white storm has passed, let it be the women, then, of your tribe who in the

Moon When the First Snow Flies gather to determine by a process of equal vote who are to be your principal chiefs and leaders for the coming season."

At this there was an audible rumble of surprise.

The concept of voting was not itself new. Fundamentally that is how chiefs and leaders had always been determined. However, in the past, tribal voting sessions had always been informal affairs, dominated by the most vocal of the men. The women had played indirect roles by helping to form the opinions of their brothers, husbands and fathers. But it was not often that any but the stronger and more outspoken women influenced the process directly. To give women not only a role but the central role in a more formalized and ritualistic process of voting was a new concept. It was welcomed in a spirit of good will that soon settled into reverent deliberation.

The next few days were spent finalizing the structure of the new League of Iroquois Nations.

The confederacy was divided into three bodies: the Mohawk, *Keepers of the Eastern Door,* joined with the Seneca, *Keepers of the Western Door,* to form one of these bodies; the Oneida joined with the Cayuga to form another; in the middle, both politically and geographically, were the Onondaga, *Keepers of the Central Fire.* If the two tribes of the one body agreed with the two tribes of the other on a decision, it was referred to the Keepers of the Central Fire. If the Onondaga did not agree with the proposed action, it was referred back to the four-tribe council for further deliberation. If the four tribes still agreed on the matter, it became official policy.

On those occasions when there was a lack of agreement between the Mohawk/Seneca side of the council fire and the Oneida/Cayuga side of the council fire, the matter was referred to the Fire Keepers for a final decision. In the end there was always agreement.

During all deliberations, open debate was encouraged among all fifty of the tribal leaders—or *sachems*—ten of whom were elected to represent the people of each of the five founding tribes. However, once a decision became the official policy of the League, the sachems made it a matter of strict honor to put aside their personal viewpoints and fully support the final decision. This principle of putting aside individual viewpoints and fully supporting the goals of the united body of nations, once they were established, was carefully adhered to throughout the early centuries of the confederacy's existence. It was a sacred and respected component of *Gayaneshakgowa*, The Great Law of Peace.

Democracy was practiced among the Iroquois Confederacy for many centuries before it was modified and adapted by the troubled American colonists who had lost patience with British royalty. Though the practice had degenerated somewhat by the 18th century, there remained then—as there does to this day in the heart of every Iroquois—a strong sense of the danger of leadership by entrenched vertical hierarchy.

It is not surprising that a young colonial boy growing up in the forests of Virginia would become familiar with these ideas, not surprising that Jefferson, Washington, Franklin and Monroe came to think as they did. Though we were not often incarnate for more than a summer or winter or two in the Americans of colonial times, always we were there. From within every tree, from the blossoms of every flower, from the eyes of birds and children, we watched these colonists. It was we who gave our venison for their tables, we who influenced their dreams, and we who through Jefferson and others assisted in the firm and absolute rejection of aristocracy on American soil.

We maintained a strong and focused presence in the American colonies of those decades and assisted the newly united states in modeling much of their government after the Iroquois Confederacy. We helped them to establish a democracy that required the periodic replacement of leaders, the recognition of certain basic and inalienable rights of its citizens, and a three-way balance of power among its elected representatives.

"We knew that these former colonists were creating the context for the great melting pot that America was soon to become. We knew that they were creating the political structure that would guide the interaction of strategic representatives of the four races during the last and most crucial centuries before the advent of the New Order of the Ages. We did not return to our more general pursuits until the new government "of the people, by the people and for the people," had been firmly established.

The Great White Roots of Peace, as they were perceived by the five nations of that now-distant day, represented a deep and genuine understanding becoming firmly rooted in the consciousness of an historical people—an understanding that has never left America and never will, though it has become distorted during the recent five-century storm.

We invite all who are able to look back at this day when the flowering tree teachings were reestablished in America, those of you who can read such things in the records left in the electron spin of the leaves of trees that today grow in those same places, those of you who can cast your minds back to read the records of this autumnal gathering on Onondaga's shore, you sensitives, poets. We invite you to explore these times that we of the spirit world can only hint at in the short pages of this volume. What transpired there was in seed form what you are now experiencing as a global spiritual awakening.

The winds of consciousness that blow through those at the leading edge of human thought in the late 20th century are the same winds that rippled the leaves upon the trees around Hiawatha as he spoke, the same winds that drew the smoke from the many little fires of that autumn council and brought them together in a single ascending spiral that to this day continues to bless the heavens of human awareness. The account that we offer in the leaves of this book gives but a pencil sketch of the events that occurred. Our words are but symbols, etchings, rough outlines. The reality behind them lives and breathes. You will find it in your heart. It is for you to bring to the world of today.

The smoke that rose from those fires that once burned on the hills of upper New York State at the founding of the Iroquois League continues to rise today. It is rising in the awareness of decision-makers in a United Nations, centered not far from where the Tree of the Great Peace was first planted. It is rising as new ideas, new concepts, new insights into the nature of peace, rising in states around the world whose boundaries are slowly melting, whose aristocrats are, one by one, stepping down and coming to sit quietly in the company of peers, among circles of friends, equals beneath the boughs of a new tree flowering.

Many smell the smoke of our fires. Thought fires they are. Kindled in love. Burning for you. Calling. Calling you home.

11

The New
Information Society

*S*o it was in the Moon When
the Geese Fly South, in the year of the planting of the Tree
of the Great Long Leaves, in the century that was the
eleventh to follow our activities in Galilee, that peace came
to the North American forests. In time, the Tuscarora joined
the Confederacy, completing the body of Six Nations as it is
historically remembered. As the years passed, dozens of
other tribes and nations came to enjoy the peace that spread
out from this powerful alliance. Influencing tribes as far
away as the Osage and the Omaha, the Iroquois League did
much to spread the wisdom of the Sacred Hoop and the
peace that follows it just as surely as summer follows
spring.

For nearly three centuries, the religion of the circle was
honored again among the Iroquois and among the many

tribes with whom they traded. The teachings of the flowering tree came to influence those who dwelt in the great central valley of the Mississippi, the Ohio river valley, the Montreal and Tennessee river valleys and all the forests in between. The deities of sky, sun, moon and rainbow were appreciated and celebrated. Weapons were few.

From Lake Onondaga peaceful trade spread for hundreds of miles to the north, south, east and west. Beautiful artwork was created: jewelry, pottery, paintings of exquisite and insightful design. Great mounds of earth in the shape of animals and birds were uncovered and once again became centers of spiritual unity and harmony among the nations. Ceremonies among circles of stone reminded people that life itself was the greatest of all ceremonies.

Though it was the coming of the Europeans that finally obliterated the last vestiges of the Great Peace, by then the League had already degenerated. Scanning over the images the present American generation holds of this era, we see numerous misperceptions. On one hand there are images from those who have romanticized the native Americans of the 15th and 16th centuries, depicting them as an idyllic people, which they certainly at that time were not. On the other hand, there are the "official images" of which virtually every American school child is familiar. Neither of these extremes is accurate. Yet the truth lies closer to the images of the romanticists, who at least had some sense of the vision world that the native Americans inhabited—a world of consciousness that to this day Europeans have not understood. Intuitively, the more perceptive of the romanticists were able to sense the state of American affairs as they did indeed exist *thousands of years before* the European conquest, and in good conscience, they confused this earlier time with the America of New Amsterdam, fur trades, Pilgrims and Champlain.

Even after most of us had retreated into the spirit world, even in locations where none of us had been incarnate for a thousand years, native American tribes continued to at least recognize our spirit presence. In South and Central America, values degenerated more rapidly, but in North America—right up until the coming of the white man—there were basic values that were never violated.

For example, until the Europeans introduced it as a genocidal measure to eliminate the natives, the concept of scalping was unheard of. The concept of hunting and killing an animal solely for its fur was equally foreign to our people. Even among the most fearsome American tribes in Pre-Columbian times, an animal was never killed without giving thanks to its spirit, without utilizing its meat in reverence and gratitude.

But the cold merchants who were allowed to prosper in the cults of European monarchs during the 16th and early 17th centuries saw the natives of America as tools, dupes, fools, to be bribed into their employ when convenient and to be hunted as animals when exterminating them would make for greater profits. When these merchants had their academicians write the history books, they made no mention of the vision world or nobler characteristics of native American people. Instead they made much of the doings of the more unscrupulous natives, depicting all Americans as savages and heathen. But this was to be expected. It was, after all, still the dark ages. Only the very worst—or the very finest—of America's native children continued with their walks upon the earth. The gentler spirits had long before chosen to withhold incarnation. By the middle of the

17th century, even natives of warrior inclination began to do the same. Between 1650 and 1900 the aboriginal population of North America was reduced by nearly eighty percent.

Neither Hiawatha nor Deganawida could turn back the tide of history. Even the Iroquois League could not hold back the inevitable mingling of the four races that was destined to precede our return to the earth. But just as for Europeans the teachings of Jesus helped to diminish the violence of the past two millennia, so did our activities at the founding of the Iroquois League and the founding of the present American system of government help to diminish the violence of these more recent centuries. You remember these years of *anno Domini* as violent times, but the fact that you remember them at all—and have preserved records and delicate artifacts—is evidence that they were a vast improvement upon the darkness that preceded them.

As you thank God for the people and ideas that helped to make these centuries much better than the worst they could have been, thank also the deities of sky, sun, moon and star. Thank the deities that live within the trees, that bubble forth in the springs that give you your water to drink, that ripen the wheat, the corn, the potato in the field. For all of these deities, ignored though they have been, have continued to exercise a benevolent influence and have helped to lighten for you the trauma of these past centuries. Speaking in whispers here and there, as gentle voices in your dreams, as soothing images from the trees outside your window as you drift at night into sleep, in a myriad of unnoticed ways, they have come to you and taken the edge off your aggressiveness. Thank them. They have forestalled many sorrows. Honor and revere them. They are the servants of God.

Though these are the same deities the Tuatha De Danann honored on the shores of Killarney, the same deities that the druids thanked with their first cup of water at every bubbling forest spring, the same beings the African bushmen and Australian aborigines call upon today, they are not "pagan" gods. They are helpful beings, essential energies, not one of which lacks consciousness. They are specific ways that God has chosen to serve you in this material world. Every time you give reverence to any one of these, you honor God.

You are perhaps the only people that has not a name for these subtle essences of divinity. Forming the great pantheon of wonder, vibrating and scintillating like the colors of the rainbow, they unite in a single Great Spirit, the same Christ, the same Wakan Tanka, the same eternal starfire love, the spark everywhere honored as the source of life.

Just as there is no conflict in reality between the elemental forces of the earth and the whisperings of the solar wind, just as there is no conflict between ocean and shore, between female and male, between moonlight, sunlight, noontide or night, neither is there conflict between the monotheism of the European traditions and the pantheism of earth peoples the world over. Let the distortions and abuses of both traditions fall away.

In this new world of light, see God in all things. Understand that to honor any aspect of God is to honor God. Let reverence be not reserved solely for an image in your cathedral—or consciousness. Extend it to every sparrow, every child, every flower. Give it to every pebble sparkling on the beach.

Honor the spirit of God, not in the abstract, but in your sisters, in your brothers, and in your own heart. Give freely of your reverence. Give it to the deer of the forest, to the eagle who winds her way through the mountain passes, to the beaver who loves every moment of his life. This is how you honor God, by honoring all created things. Give thanks to these many beings who have made this age so much brighter for you. Give them your love. See them in every shaft of moonlight and in every forest glade.

When you know the Christ whose body all the earth reveals, when God's light lives within you, when your individuality has humbly offered all that it is to the service of your universality and you know yourself as one with the great Being whose representative on earth you are, then you will see a pantheon of deities indeed! Each of them singing to you of the joy, the wonder, the beautiful and blessed glory of the One.

Living in the One, in tune with the One, sing your individuality in harmony with the others of your nation, as your nation sings in harmony with the other nations that in this world are forming the organ of consciousness of the One. You are the way that God becomes human, the way that God tends the garden. You are the way God interacts with the universe of creation. This world is your unfoldment. You have created it that as this human family you might come to enjoy time through the substance of biology.

It is odd for us now to address you in this way, we, that part of yourself that stayed awake while sleeping, you drew this species out of your dreams; for you incarnate humans of this age slumber yet in the image of our children. Yet we know we awaken you best through speaking to you as our source. How can the parent speak to the child, knowing that soon the child will become a God? How can we who have guided your sleepwalking steps—and saved you time and time again from tottering over the cliffs of ignorance—now shift perspective and perceive you as we did when the world was young and your Spirit gave us birth? How? With pleasure. With joy. And with song forever ascending in the upward spiraling fires of your eternal love.

With all its high adventure, with all its pageantry, with its epics, both recorded and thankfully forgotten, this fall of your human consciousness is, of course, healed, as all things are in the passage of time. Intelligence once implanted, though it can be misdirected for an epoch or an age, nevertheless remains intelligence at its source and cannot, no cannot, possibly fail to learn.

In the obvious and evident simplicity of the truth that your human consciousness now understands and communicates so freely, you cannot help but wonder why these ideas could not have spread as easily among Socrates' schools, in Renaissance Florence, in the courts of Solomon, or in the monasteries of Tibet. For a fleeting moment or two as the dawn streaks blue and brilliant on the horizon, you wonder.

Why did the truth not flower when Columba cried on Iona's shore? Why did the lightning between Francis of Assisi and Pope Innocent III illuminate only the Vatican? And there but for an hour? Why did it fail to ignite a fire that would brighten the world? You wonder why Arthur, in the end, could not hold the table round; and why the finest Nordic generation yielded to the hypnosis that taught them as much as the Japanese.

You were there then, in the tea gardens of the Emperor, when it was truly a Sun God who sought to fill the minds and hearts of men. And was that not you on your horse, riding into the desert, disgusted with the warrior ways of

China, who stopped for a moment to pen those words, which have been long revered, but which have never yet accomplished their goal? And under the bodhi tree, after you had eaten the milk and curd, why did you ignore your impulse to speak at length to the woman who offered it into your hands?

In this moment when history is about to be laid to rest, in this flickering moment before your attention turns toward the stars, you cannot help but wonder at those critical junctures. The truth is spreading so obviously and naturally among the humans of the late 20th century. Why did it not spread so easily and naturally when perhaps you saw the chance, incarnate here and there in other lives, in other moments? Long have you passed in and out of these realms, holding candles in the darkness, drawing the warrior tribes closer and closer to the healing that is now no candle, but dawn flaming brilliant across the eastern sky.

Today you see these ideas appearing in hundreds of new books, songs, movies, publications and periodicals. As you see the passion with which they are translated into every human tongue, converted into every form of media and shared among the healers and leaders of every nation on earth, you wonder at the moments missed.

But then you see that there were no moments missed. And that had you followed the woman who offered you the milk and curd, the two of you together, united in bliss, might have taken her canoe far beyond the Ganges and communicated not a word to the illusion-bound. As you did in the end. Quetzalcoatl, Jesus, you know how they were beginning to depend on you. The truth did not diminish when your human focus did. It grew. Deepened.

Though the Great White Roots of Peace care not for names, egos, labels, images that rest like yesterday's leaves beneath the pulsing of this season's life, nevertheless, all are honored as every root of peace weaves together in a great concentric tree-ring network of light that runs through the first great Mother and flows through the blood of every one who has ever lived to make this a better world.

Suddenly you see that there is a reason why it is different now. For when you formed the Tao Te Ching, you sat alone in the sand, your horse grazing on what grass there was at desert's edge; there was only one to take your written words. No telegraph existed to relay them with lightning speed to the other side of the world.

And when you spoke of the Eightfold Path, a few around you understood, but their articulation was not yours, and no recordings could preserve the truth more clearly. And when Francis spoke, no torch of video carried that lightning illumination to kindle fires throughout the far-flung reaches of Christendom, though kindling was gathered on many shores and people prayed for just that flame.

And when the spirit filled Jefferson those weeks of that New England year, there were moments, you know, as Thomas spoke, when the spirits of God looked out of the eyes of many a former Englishman. Yet that event was not televised, beamed through a global telecommunication network to those who watched from Parliament and homes on every continent.

There are many reasons why consciousness is presently spreading so rapidly: the tremendous explosion in human connectivity brought about by telephone, radio, television and orbital satellite relay stations; the exponential explosion of information due to the tape recorder, video recorder, film, and computer technology.

No sooner are the words spoken by the White Buffalo Calf Women, the Deganawidas and Hiawathas of today but they echo through the heavens, flow through cables that run beneath the ocean floor, sail tower to tower, leapfrogging microwave across the country. Simultaneously, they whistle in telephone wires across Nebraska and the Ukraine. They sail, electronic signals at the speed of light—satellite to earth, to satellite and to earth again, circling the globe. Even as the words are spoken, the ears of the world are listening, waiting, open to receive.

But the increase in human connectivity resulting from communication technology is less significant in the present global awakening than the sheer numbers of people who are communicating what Francis communicated in medieval Europe, what Lao Tsu stated at the edge of China, what Hiawatha shared on the shores of Onondaga and Jesus taught in the streets of Jerusalem.

Our company in those former times was small. So rigid were the preconceptions of the collective human consciousness that we had to enter, it was often the case that it took our entire Light Circle working together to help sustain an incarnation in even just one, let alone two or three who could speak the truth and bring the light.

Not until these closing years of the 20th century have we been able to incarnate fully and completely, aligning our spiritual bodies with the physical, etheric, mental and emotional bodies of not just one woman or one man here or there, but with many, at times even whole communities.

As the new information society bubbles over into global spiritual awakening, we come to join with our human egos on a scale unprecedented throughout all of history. People turn once again to the Great Spirit who is the source of all information. Rapidly in some, gradually in others, our egos discover their appropriate roles in the communication of this new awareness. And there is naturally distortion at first. But in the explosion of consciousness, it is soon purified. Our spirits incarnate in ever-growing numbers. Distortion diminishes. Truth spreads. And how beautiful it is!

Like crocuses announcing the coming of spring, channeling proliferates for a season. Then, just as crocuses give way to daffodils, channeling in season gives way to a deeper and fuller communion. The spirits of God secure the cooperation of the egos with whom they commune on earth. Slowly and steadily they proceed through the process of bonding with their egos, no longer just speaking to them, messengers from the stars, but flowering into perfect union, waking up within them as the beings they themselves are.

And on it goes. Daffodils follow crocuses, dandelions follow daffodils. Dogwood. Apple. Pear. Plum. All blossom. Oak shakes her gentle flowers and summer comes to the world at last, watching all teachings fall and pass behind her as the flowers and the early blossoms of spring give way to the unabashed sensuality of midsummer. Luxuriant. Rich. Magnificent.

There is gratitude for these teachings that helped to turn human hearts from violence during the darkened times. But one does not eat out of a garden that grew in centuries past. Moment by moment, the Great White Roots of Peace birth the same truths fresh, new again. Like this summer's vegetables. Like this season's fruits. To be enjoyed, appreciated and shared when ripe, to be cared for, that they might continue to blossom and flower. Like the Ongwhehonwhe. Like the people true to reality. Like the Bird Tribes whose seeds have lain dormant in human hearts for thousands of patient circlings of this earth around her star.

12

The Great Day of Purification

Listen, O humans of this present world; listen, as the sparrow listens for her lover's call on the breeze. We are speaking to you in your dreams, in snatches of song heard in passing, from the mouths of children, from these pages; but more than this, we are speaking to you from the center of your innermost being. Hear us, and remember yourselves.

We return to human consciousness, bringing a time of new creation and the information required for humans to understand the changing conditions of the age. We are here to merge, to blend with your human egos, to help your race become the central guidance system of a vast new being. We are here to help the Great Spirit incarnate in the people of the four races.

We have come to initiate the most joyous age this earth has ever known: a thousand-year period of earth healing and renewed harmony that will see the four-legged flourish, the two-legged awaken, the rivers run pure and humankind in conscious exchange with the stars.

And after this millennium has in turning season passed, you—a new being of translucent biology—will depart for the vast, uncharted ocean of the Milky Way to give the freedom of the Thunderbird's flight as a gift to every star and every world. In a multiplication of love affairs that will go on until the end of time, you and your race will seed the galaxies with intelligent, biological life.

The potential of this universe is limitless. Her revelations of potential are infinite beyond number. But there are no star wars or advanced and fearsome civilizations beyond your own; because if civilizations are fearsome, they do not advance beyond your own. They become extinct, to rise up in the soils of another world, a little wiser, until they learn.

Our task today is to help you remove the blinders that historically have distorted your perception. Realize that what you feel in your heart determines what you see. Perception rides upon the expressions of the heart like a canoe rides upon the waters. When your heart expresses fear in any of its turbulent forms, your understanding becomes jumbled, confused, you perceive through the waves of illusion. But when you love, you understand, for then you share the vision, the very perception of God.

Return, like the salmon, to the place of your origin. Birth your moments only in love. You can root your life in fear and know the predictability of granite, the strength of marble and what security there is in limestone's patient changing. Or you can root your life in love and join us in helping to ease the human world through the awesome changes of these times.

The Great Day of Purification has begun, a short but essential cycle of division that will gather those who promote fear and violence and separate them from this season of the world as chaff is separated from wheat at threshing.

During these next two and a half decades, humankind will experience this separation, a transformation in human consciousness more fundamental than the development of language, more significant than the shift to agriculture, more meaningful than any historical revolution. A new era is dawning.

As the sun rises, the shadows become sharper and more clearly defined. As the energies of love grow stronger, the shadows of fear become more visible than before. To some it may even appear that they have grown in number and in strength, but this is not so. What was hidden has simply become revealed—that it might be healed and brought to peace.

Action born from fear is becoming less effective with each passing day. Institutions, traditions and societies forged in fear have already begun to falter. Change is occurring everywhere, for as the planetary awakening proceeds, the consciousness that determines the quality and nature of life is itself changing.

Amidst all these changes, more and more people are turning to God for guidance and direction. They are coming to dwell in the place of their own inner spirits and to recognize their spirits as expressions of God. That is where God asks to be trusted—in the human heart. That is where the awakening takes place. The Creator asks not to be worshipped in an external image, but to be acknowledged in each human being.

You are all god beings *in potential* with no reason not to become God beings in reality. Trust yourself, trust your natural response to each new situation. The action arising from within your heart is not going to be destructive, it is going to suggest the most creative path to walk in answer to your situation and your world. When you trust yourself, you are trusting in the Wisdom that designed you. This is how you trust in God. It is not an abstract thing.

Trust in God is trusting in the God who lives within you, trusting in your spirit's ability to respond to each situation beautifully, impeccably, individually, creatively. When you doubt your native ability to breathe the air of spirit into your world and create according to your divine thought, you are doubting both God and the universe. You are rejecting life's most precious gift to you—your own inner knowing—and you are presuming to replace it with values, judgments and opinions you have acquired secondhand.

Without the acquisition of another skill, without the acquisition of anything but complete and total trust in God and in yourself, you have everything you need to interact optimally, creatively and productively with every situation you encounter. *There is no exception.*

Perhaps as much as seventy to eighty percent of your current typical behavior is exactly what you would do in a fully awakened state. Accept yourself, feel clear about *all* you do. As you do this, you begin to introduce into your actions the very power of the Creator.

There is a movement of spirit that proceeds from within your heart to greet the world with the clarity of perfect action. That movement is intuitive. It is your direct link with the Source of all life. In the instant you know what to do, and in the same instant, you flow into the perfect action required. Trust your intuition. It is an arrow whose shaft has been carved straight and smooth; unerring and true it flies to its mark. Can the ponderous tread of the rational mind be compared to the swift, sure flight of an arrow? Reason is designed to support, not to lead your action. It is meant to help you implement the purposes of your heart; it is not meant to determine them.

Trust in God by honoring and trusting your intuitive sense. No God would create a creature without the wisdom to chart its course! Trust the nature of your design. God's actions appear within your awareness as the most natural thing for you to do. Following them will reconnect you with the awesome powers of the universe, for all your actions will then be in harmony with the underlying intent of the life force itself.

The incarnational process is a process of relaxation. It is a process of relaxing all thinking that is born from straining and holding on. It is a process of choosing not to struggle against the current of what comes to you through the natural movement of your heart. This does not mean that you close your eyes to the world around you and try to "think only good thoughts." No, you can look at the hunger in Africa and do something about it. You can look at the need for greater communication between nations and do something about it. You can perceive needs in any area and take action, motivated by love, to alleviate the suffering in that area. But though they may encounter problem areas, when your thoughts spring from intuition, they are not sustained through tension, anxiety or fear. They are characterized by

God's love, flowing like powerful energy through your understanding. Happy, knowing that you are there; grateful, because you know you *can* make a difference.

Invite the Bird Tribes into your awareness. Make a home for eternal spirit in your heart. Notice what thoughts create tension or anxiety and choose to release them. To the degree that anxiety taints your motivation, to that same degree, you are less effective in all you do, for spirit is then denied access to your mind and heart, and *you*, in a literal sense, are not fully present.

The process of relaxing all thoughts that require tension to sustain them is a process of relaxing habits of the past and awakening to the reality of your own spiritual presence. It is a process of losing interest in fearful ways of walking upon the earth and choosing to create your world in love through the action of your open heart.

Should you perceive a problem, you surround that problem with love. You define it in love. You recognize that it is caused by the lack of love that lies behind all problems. And with love you introduce the understanding that provides the solution.

On the incarnational journey, you lose interest in all that perpetuates this rapidly passing state of illusion on earth and you become keenly interested in what is in the process of replacing it. You begin to perceive an emerging Sacred Reality.

In the whisper of the pines, in the gentle call of the morning dove, in a child's sleepy yawn, the new world comes dawning. Through the facade of the late 20th century human world it shines: a world of shimmering potential, a world of beautiful light, a world where human beings do not behave violently toward one another in thought, word, or deed. It is a world of understanding and cooperation, a world of abundance in which the human spirit, freed from the shackles of self-centeredness, flies on the wings of love, leaving a trail of joy, wonder and undreamt-of beauty wherever it passes.

You do not have to be a scholar to notice the direction and momentum of history. Even a child can observe these times and know that fundamental changes are being wrought deep in the collective heart of humankind. Something is changing on the surface as well. *You are here to help it change easily.*

You have only to make the choice to shift the center of your motivation from fear to love. Make the choice. Begin to identify the working of fear in your life and recognize it for what it is. There are certain human fears that may have small supportive roles to play in the course of this transition—if you stepped out in front of a speeding automobile, for example, you would be grateful for the fear that would cause you to jump back out of its way. But fear's intended role is a small one. Compared to the historical state in which fear has been the keeper of your power, fear's true role is a small one indeed.

When you choose to relax, you break away from the control of fear's conditioning. You break the bonds that have historically enslaved your race. In the expression and restful enjoyment of love, you come to know the energies of the Great Spirit.

Pouring forth from the heart of the Nagual are powerful streams of energy, which we of the angelic tribes channel into the emotional field around the earth. These energies are designed for creation, but when they meet with disharmony, as is the case with your present-day human world, they translate into energies of healing, forgiveness and education. We can direct these energies in the general sense, but before they can reach their full power, they must be consciously directed by people who are awake and incarnate on earth. Only then do they enter the realms of specific application that will release their fullness and, during the remainder of this cycle, complete the healing of your race.

You are being invited to open your heart and make a welcome for these immense creative energies. You are being invited to help direct the powers of eternal love.

There are two requirements. The first is that your heart be open, loving and able to channel, at least to some degree, the love of God. The second requirement is that your identity be fully in the present moment. If your identity is based in your past experiences, *you* are not truly present. What looks through your eyes is then only a fictitious creature, an image, an illusion.

Too much thinking about oneself is the greatest thing that keeps human identity from being fully present, for when you are constantly self-reflecting you are too caught up in past and future to notice the presence around you. You are doubting your own power. You are not vibrating fast enough to channel the immense energies of creation because your attention is scattered and you are closed to the one moment where the love that would quicken you exists: the moment where you are.

There is nothing wrong with human individuality; it is as necessary to God as leaves are to a tree. But historically you have made a god of leaves and have forgotten the tree at the source of your life. If you truly wish to channel God's blessings, begin thinking of yourself less as an isolated self and more as one of a family. You are *all* of human society; *you are the world* and all within it. When you relax into perception of *other as self,* when you think of yourself as the pair of eyes into which you look, when you care about those around you as much as you once cared about yourself, then the energies of healing and transformation are able to flow freely through you. Your eyes, your ears, your every sense is attuned to the richness of the surrounding world. All your senses are operating at a rapid enough level of perception, correction, and decision making to keep up with the energies pouring into your heart from the Great Spirit who comes to fill the needs of the world through you.

Love vibrates rapidly. Fear has a slower rate of vibration. Those who channel fearful energies find that as time passes, the fear vibration grows heavy, depressing. Eventually, it brings sleep, gloominess, discouragement, despair. The love vibration brings enthusiasm, energy, interest, perception. And that is what your world needs in this age: your love *and your perception.* This is what will heal the world: clear and undistorted perception, flowing through a you that is not self-reflective in the egoic sense, but self-reflective in the sense of knowing the god within.

The self to reflect on is the self that you truly are—the Creator, the Eternal One, the spirit of God. You reflect on those aspects of the Great Spirit that manifest before you as the men, the women, the children, the plants, the animals, the crystals, the creatures of the world entrusted to your care. First and foremost you are a steward of your immediate world. You care for the earth, not in some nebulous way, but *directly* as an individual representative of God. You take responsibility for bringing love and understanding into every environment through which you pass.

As the birds welcome the morning sun with their song, when you act totally in love, you welcome the Great Spirit into your world. You become a conscious cell in the awakening earth. The physical atoms of your body sing together, vibrating in perfect harmony. Your integrity and unity of being, your integration of purpose and expression brings resonance to everything you are and to everything you perceive. You become a natural channel for the vast and powerful energies of creation. Through you they flow into the healing and transformative work of these decades.

Pulsing with the atoms of your body, pulsing in perfect rhythm with the world around you, you know yourself as a localized expression of your own universal being. You know your *First Self* as the Great Spirit who is awakening within the human populations of the earth. And, as a son or daughter of the natural world, you know your *Second Self* as an individual human being within those populations.

As one who has awakened the Bird Tribe energy, your eyes behold the world around you as the divine world within. *And you know you are inside.* You, the Creator and the world are one. You are awake. Your biosphere is active and alive. Just as bees are in communication with the spirit of their hive, you know in an instant all that is transpiring throughout the field of collective human consciousness. From the collective field of your larger awareness of the earth, you are able to instantaneously access the knowledge you need in the moment. You use your individual free will to pursue God's will—and you know God's will as your own. Your identity rides the crest of unity's wave, at the very meeting place of Creator and creation.

You are free, immersed in universal currents of love. The Great Spirit's own consciousness flows into your heart and perceives all through you. This is the gift to the innocent: *to see through the eyes of God.*

PART TWO

The Serpentine Mind
Disentangles
From the branches of confusion
Uncoils its knowledge to greet the dawn
And sees
In the growing light

Great colored, feathered wings
That grew unnoticed in its sleep

Understanding perfectly
In a single rush of joy

It flies above the ancient promises
Beyond where the forest lies
To the sea
The open sea

As rings around the Morning Star
Light corridors through the air

—Quetzalcoatl

13

The Energy
Behind the Wind

*B*ehind the subtle breezes that whisper through the treetops, fine and delicate energies pulse into your world from the spirit world, from an energy level not visible to your senses but as tangible as all you touch and hear. You detect these energies, taste them, feel them. They come pulsing out of the early morning air above the trees, gliding swiftly past you, lifting tiny hairs on the surface of your skin. Out into the meadow and beyond they flow. You cannot see them, but you sense them and know they are there.

Watching these subtle whispers behind the wind, you notice that the birds are watching them too, playing with them, learning from them the news, gathering a picture of what the day has in store. You become aware of little messages going back and forth throughout the forest. Every tree, every fern, every moss-covered stone is aware, sensitive, listening.

The natural world is attuned to a multidimensional communication that is always happening—flowing back and forth. As you listen to it, the wind tells you things. You allow its currents to gently move your thoughts as the eagle flying low over the treetops lets them move his wings. You sense how the eagle waited in the tallest tree on the ridge to catch a certain expected current, riding it when it came, gliding before it as it broke. Like a wave, releasing warm air across the valley.

If there were words put to these subtle conversations of nature, what message would they convey, these designs and patterns, sliding quietly into your world from a realm beyond? You watch and listen, blending with what comes your way. Somewhere in these currents, there is meaning for you.

You notice that these energies, more subtle than the breezes, are in fact the reason why the winds circle as they do. The currents of moving air respond to these subtler energies and express a portion of their information flow. You begin to see in a new way: the forest, a system of living information flowing back and forth along an invisible, but to you, an increasingly perceptible network of energy. You feel it, picture it, with something deeper and more clear-sighted than the physical eye: little webs of pulsing current, crisscrossing among the trees, networking, joining, slowly circling in the swirling presence of a vast and eternal being. And along these webs comes information. Sailing. From one world to another.

As you relax, you blend into this pulsing network. You no longer feel separate from the grid of its swirling creative energy. You feel it within you, around you. Merging back into the ground of being, the little voices behind the wind lead you into an awareness of the *intention* that is calling out the life of this earth. You release all tension that would hold and define that great love. You relax your self images, your ideas, your beliefs. You let go of your concepts about being human. As your cultural definitions fall away, you experience a natural expansion of awareness. You know yourself as a part of the world around you. It is a physical as well as a spiritual experience. You no longer force the creative energy of life into arbitrary structures of interpretation. You let the energy flow as it will. You become aware of the living presence of God.

You begin to see, as you blend with the subtle energies behind each breath of wind, that the sun's influence contains information as well as warmth and light. Warmth: life information. Light: intelligent design.

As you merge into the ground of being, allowing your sense of self to be drawn into an expansive new experience of who you are, you encounter the solar intelligence behind the subtle breezes. You hear a familiar voice, like the voice that whispers in your dreams. It tells you things, things that at first come to you as matter-of-fact observations, things that, of course, you had always known. And yet. And yet, when in deepest relaxation, you allow these solar winds to blow through the words you have stored in human mind, you realize that these insights, obvious to you in your natural state, are not acknowledged by the human world in which you live—though children and poets at times sense them still.

You listen to catch the words that will help translate your perception.

Information percolates through to you from the subtle world of spirit. Your body knows, your senses pick it up. Its relevance goes deep into that place in your heart where you determine how you can make the most creative use of time.

Every dawn brings indication of activities for which the coming day is suited. When you saw the eagle fly across the face of the rising sun and give out his call, it was not accidental or without meaning. To the other birds and animals who were watching and gaining their impressions of the new day, the eagle's flight was a cue, a message rich with meaning.

The time that he chose for flight, the direction he was flying, the place from which he appeared to the birds and animals who were watching—these were parts of his statement, sentences in the paragraph of his expression. But there was other news, equally valid, of more or less importance to various creatures. How the crows reacted to the eagle's flight. How long after his cry the owl and whippoorwill chose to pause before resuming their song.

The birds and animals notice. What kind of day is it going to be? Is it a day that you should look for your meal on the hillside? Or might you be better off going up into the pasture? Could it be a morning to fly to the treetops and sing for an hour more? Or is it a morning when you had better get busy, because it might cloud up and rain before the day is over?

In every moment the Great Spirit communicates to all creatures everything they need to know. Through ten thousand billion agents—angel, elemental, animal, vegetable and mineral—through the vast and subtle network of living design beyond the weather, before the wind, the truth is ever being transmitted into this world of form.

It is up to each one to sense how that truth translates and relates to him or her in each moment of the day. And this is as true for humans as it is for any other creature. The process is not complex, the sensing and translating is not done with the mind. It is an autonomic process that occurs spontaneously below the level of thought when judgment subsides and allows perception to simply be, a natural process that takes place effortlessly when your mind relaxes its cultural interpretations and trusts *you* to experience the natural clarity that is always present—*when you are present.*

In the historical state this natural process is overridden by human insistence on making decisions with the intellect alone.

Intellect is a good thing; reason is a valuable tool. But the mind was created to serve, not to eclipse the human spirit. In attunement with the creative energies that pulse beneath the outer surface of life, spirit has ways of determining behavior that are far more rapid and effective than anything an ego could calculate through slow-motion, linear reasoning.

Notice first from below the level of your reasoning, from above the level of your individual identity. Notice the rich and awesome world around you. Feel its impressions. Notice how they change like a kaleidoscope slowly rotating, shifting and rearranging colors, hues, shapes and vibrations as they tumble slowly over one moment into the next. In every moment you have around you all the information you require. You are at all times surrounded by *the truth of what is*, the energies behind the wind. They are signaling. To you.

On the subtle levels of being that precede the winds, the spirits who take form on earth first enter this material plane. Most of them enter on the solar wind and live only for a moment or two, an hour or a day. Ice crystal egos form around them in the upper atmosphere and melt before they touch the ground. But some learn to live longer. Some find their way into feather and fur.

What comes before the morning breeze? Subtle intentions of the solar wind. What comes before the breath of sun that comes before the wind? The energies of being. On a planet. Turning. To face her love-star.

Earth and sun. Building a spirit bridge along the lines of attention that they turn one toward the other. Lovers gazing into one another's eyes.

Along the lines of their gaze, potential appears. Some of that potential becomes incarnate, most does not. Some succeeds in attracting physical form, some does not attract even that first crystal. An invisible field of energy around the rolling earth, a *vibrational grid of form in essence* determines which patterns most easily materialize. A grid of intended creation is held suspended around the world—an image of what is to be. An order. Implicate. Becoming.

You feel yourself relaxing into deep awareness of our kind. You feel us surrounding you, working with you, assisting you in the world you have chosen.

Ideas occur along the lines of gravitation between earth and star. We hold some of these ideas as one might hold an image in the mind. We visualize what we would like to see incarnate. Eventually creation will reflect our designs, for along the energy pattern images we suspend in consciousness, creatures take form.

The creatures that find themselves incarnate in biological form, in feather, leaf or flesh, have formed along the lines of this energy grid that exists on the surface of the earth where her atmosphere mingles with the light of this nearest star. This energy grid is our perfect image of a child, a child implicit in the order of the universe. The child in which the spirit of the Eternal One begins to awaken. This is the being of God, drawing the earth's biosphere into a single living body.

Everything in nature pulses with a sense of this Being, every ripple on the ocean, every hill and every mountain. Every particle of matter contains the total information of all creation. It is radiating out from everything. Relaxing your individual sense of self, this is the Being that you become. Totally satisfying. What your heart has been longing for. Releasing fictitious images to the wind, you feel, you know. You accept the image and likeness of God, the perfect pattern image of the angel that you are. In that acceptance, you experience the fulfillment that is your mission and purpose in time.

There is a perfect image for you, a spirit energy field that calls your biology into the vibrating patterns of Being's perfect purpose. You have only to relax and release the ways you have formerly thought of yourself to sense this perfect image, to feel your spirit purpose, to allow God to describe you.

Why create yourself in the image of imperfection, when the Great Spirit, source of all life, holds out to you an image of unspeakable beauty? Can you feel it now, fluttering into the atmosphere of your consciousness? It is defining you as a projection of all that is, as a steward, a caretaker, an Ongwhehonwhe, bearing special gifts of love for the earth and for all her creatures, for the fishes of the sea, for the four-footed, the winged and for all that you perceive. This world lives inside you; her creatures are the organs of your body; they are a part of you; they *are you.*

You are this world's organ of consciousness, awakening. You are eternity's love for time—embodied in human form.

Your being and God's being are one.

Noticing the world around you, perceiving the daily information of the subtle energies that precede the wind, you do not fail to note the seasons. With every fiber of your being, you sense the shifting climate. You feel the returning spirits of the stars and know well that the old world is passing.

You see children dance up out of the earth who perceive clearly the world of reality. Their bodies are self-replicating. Their minds are light. They are of the new consciousness— our flames across the shimmering lakes of understanding. Consciousness. Come from the stars. Appearing from within your own hearts, beginning the objectification of a presence long overdue on this world, Etenoha.

And the new consciousness speaks of the circle, of the round, of the Sacred Hoop, of giving and receiving, of seasons for harvest and seasons to gather up the fruits of all that has been. A new harvest, a new species, loving people to explore and develop, to enjoy every sunlight, every shade, every hillside, every field, every river valley.

As the ages pass, you will see many more worlds in the temperature range of this planet, this third planet from your sun. It is in this temperature range that the perfect balance can be found between energy and matter, spirit and structure, Creator and creation, time and eternity. It is in this temperature range that the beautiful child, the Flying Winged One, is created and sustained.

From this temperature zone the Children of Light will go forth to cultivate the star fields, introducing these temperatures wherever there are worlds eager to know the potential hidden beneath their mountains or sleeping in their sand.

You are the beginning. The beginning of how eternity reveals its potential in time. Your own historical moment is indeed the optimal moment, long predicted. Many of your kind are now opening to the angelic blending that is guiding your race from ignorance to reality.

Humankind is poised on the threshold of a fundamental species-transition. We, and the Great One whom we serve, desire this change to come gracefully. Like the opening of a flower. Or the breath of angel wings behind the wind.

14

Shadows in the Biological Mirror

We are interested in populating a material universe with creatures that express the balance between stellar and planetary polarities. We are interested in creatures that have substance to them—flesh! bone! wing! fur! scale! ear! eye! Creatures that are sensitive to sound and light. We want to inhabit forms of soil and starlight. We are interested in people: artists, creators, lovers singing the elements to life.

And so we create images of what we would like to see clothed in the worlds of form. Creative images. Entities. Sent forth to dimensionalize in full, seven-spectral vibrational matter, to clothe themselves, through paths of their own choosing, in the living biology of a loving material world.

We are both the source and the reality of these images. We are God's specific terrestrial intentions personified: highly focused angels of creation. We are the builders and the mechanics among the Bird Tribes. We fly low over the surface of the earth. We pass over her mountains and river valleys. We come to a geographical area for a reconnaissance mission, a fact finding, light-sound-sensing, exploratory mission. We fly close to gain a better idea of how we can clothe ourselves in water, air, earth and fire.

Yet no sooner do our spirits hover near her surface than we find an earth already teeming with biological suggestions. Before we have gathered even a fraction of the data we require, like prairie dust drawn up into whirlwind, the earth elements rush up to us, dancing in myriad living images saying, "Here, this is it. This is what you want. This is what you are. Here are living, three-dimensional, biological pictures of your thoughts and dreams. Foxes, fishes, kangaroos, elephants, monkeys, dinosaurs, forests, jungles, tigers, lions. Look—objectifications of the thought-designs suspended in your light fields."

While we are still observing, the earth is throwing up objectifications of every image and idea that enters our consciousness. Some of our ideas are well thought out, but builders consider many options before committing materials to a design. We are used to seeing our considerations objectified on the higher mental and spirit planes; but after examining designs on those levels, if we do not wish to sustain them, we simply turn our attention elsewhere and the objectification disappears—as if it never was. On earth it is different. The creature, whether it be a saber-toothed tiger, a butterfly, or a flower, continues to live. This is the value as well as the challenge of the earth.

"Please," we say to our elemental helpers, "slow down. We have not yet begun to create. Not consciously. Are you already that excited? We are just here to gather information."

"Let us help," reply water, air, earth and fire. "What is it you desire? Tell us. In feather, fur, scale and flesh we are giving you the information you require. We cannot help this excitement. It is what our whole world feels in the creative energies of your presence. How else do you propose to gather information? We want to help clothe you in matter. Show yourselves to us. We want to see what you look like in all your aspects. For you, we have been longing and waiting. We will do anything to help you. We love you."

"We know you mean well," we respond, "but we need more time. We would like to approach this more gradually."

And so begins our early communication with the elemental beings of the earth. Some of our communication gets through. Some of it does not. The earth forces are profoundly intelligent; they understand the structure of matter and the eternal truth it embodies more comprehensively than we do, but the nature of their intelligence is, in many respects, opposite to the nature of our own.

Intelligence is relative to the context in which it functions; it is the measure of a unit's ability to creatively interact with and within its environment. There is no such thing as purely objective intelligence. What is intelligent behavior in the cooler realms of material structure would be fatal on the surface of a star; and, conversely, what is intelligent behavior on the surface of a sun would be utterly devastating upon a material world.

Relative to the material realms, the elemental beings of the earth had reached the same high order of intelligence that we had reached in relation to our stellar realms. In that sense we each represented the pinnacle of the only two fundamental types of intelligence that had yet—up until our meeting—achieved *incarnate* expression in the universe.

What we were seeking to create upon the earth was, in effect, a whole new kind of interactive intelligence, a kind of intelligence that could, when required, embody either stellar or material intelligence, but that would place its primary value upon the synthesis of the two. Ultimately, only the new intelligence itself could determine just what that synthesis would be. This was the unknown, the variable. It rested like a gray question mark in the middle of our every equation.

Working with our elemental partners in the earth, we could guess at what values the new intelligence might prefer and approximate the conditions most likely to favor its appearance; but we could not in any way compel it to be. The new intelligence could not be manufactured; it could only grow of its own accord within a womblike environment, a world that gave it every chance to be. Working as closely with our elemental partners as the differences between us allowed, we gradually channeled more and more of the earth's biosphere into increasingly likely candidates for this new type of intelligence.

As time passed, there were critical junctures in the creative process, moments in our relationship with the elemental beings when breakthroughs occurred, when new ideas were grasped. The level of mutual creativity gradually gained precision. The earth elementals came closer and closer to understanding the creatures in which we were interested: translucent creatures capable of knowing both solar and material intelligence as their own. But what would that practically require? What kind of minerals? What kind of bone structure? What size and weight? What kind of diet? Environment? Climate? At each breakthrough of new understanding, the earth elements would catch the energizing winds of our spirit intention and sail them into increasingly more accurate embodiment.

And so the process of mutual creation and improving communication went on. For millions of years it continued. The time seemed long to the creatures of the earth, but to us, creatures seemed to come and go like successive waves of passing images splashing on a sandy shore.

So when we see a new creature rush to us out of the earth saying: "I have heard that you are looking for a creature like me. Am I the one? I will do everything I can to be the one," we do not turn that creature away, even if it is not exactly what we had in mind. We give it time to grow, learn, change and evolve—even as your own human family has done. That is the raw material from which the balanced creation is made. It takes such initiative on the part of the earth. We cannot bring about the balanced creation alone. In order for angels to be incarnate in the earth, the earth must participate as much as we do. It is only in our blending with the earth and the forces of her initiative that the stable child of time and eternity can come into being. It is only in our starfire's blending with the truth of material structure that a healthy child of spirit and matter can be born.

When the first wave of humans appeared, a breakthrough occurred in our relationship with the elemental spirits of the earth. The most significant yet. The quality and level of our spirit/matter dialogue took a quantum leap forward. Here was the connecting link, a species appearing right there in the meeting ground of our interacting values. Here at last was the unifying medium with the potential to provide optimal communication—and therefore much greater understanding—between us and our elemental partners in the earth.

Here also was a new type of challenge.

Humankind had a mind of its own.

These first human egos were our own rather poorly reflected images, semi-independent shadows in the yet unclear surface of the biological mirror, as yet without consciousness of the beings who cast them. Over the course of their development, we had shepherded them. We had cradled them, nurtured them and guided them through a long chain of experience, that this very day might at last come when we could begin the bonding that would allow them to become our conscious partners in future creation.

Of their own free will, these egos had met the universal criteria that distinguishes conscious entities: their motivation was more frequently centered around love for their fellow creatures than around fear for themselves. This choice had signaled to us that they were ready.

We could now point out to them the spirit world, introduce ourselves, and begin the final stage of the first human generation: one stellar spirit merging gracefully into each human form. The egos now had the capacity to understand the beings who had created them and the intelligence to cooperate with the purpose behind their creation.

At first our exchange was creative and fulfilling even beyond initial expectations. But as the glamour gradually wore off and gave way to the spiritual and technical mechanics of the actual bonding itself, misperceptions occurred. Some of the egos saw us as frivolous, ungrounded and lacking in respect for the material plane. Though this was not true, our own enthusiasm and incomplete understanding of material values may have contributed to this impression. Certain egos, instead of seeing us as the truth of themselves, perceived us as powerful, incomprehensible beings of light, creatures foreign, separate, apart.

We shared with them glimpses of the glorious future that would open before us as we blended into unified beings. It was too much, too fast. They reacted in fear.

History began.

Had the original design unfolded with human cooperation, it would be only after an ego had come to prefer love-motivation over fear-motivation that it would find itself assigned to a fetus that was simultaneously experiencing the entry of an angelic spirit. In healthy development, as an ego is accepted into human form, a spirit also enters the same form and bonds with the ego. The once mortal creature becomes immortal. In the bonding process, the ego learns the secrets of its origins in the earth. It remembers that spark of eternal spirit that quickened it to conception and oversaw its growth and development. With its earth identity intact, it recognizes its most essential self as a reflection of the spirit self. It accepts that spirit identity as its own. Ego and spirit bond; the polarities are connected. The mind/brain filament glows with the light of the new blended consciousness.

The incarnation is complete.

In unhealthy—or historical—development, it has worked differently. After the spirit/ego bonding was disrupted, the biologically complete (but spiritually incomplete) humans conceived their children—as they conceived their world—in fear; many egos were thereby admitted into human form who were not yet sufficiently prepared. Their motivation was still "running-from-what-they-feared" rather than "living-for-what-they-loved." These egos were intended to be the candidates, the trainees who in a healthy state would not yet be permitted human incarnations. But the first human conceptions that were not wholly loving opened the door for them to enter human form. And, of course, once that door was opened, other still fearful identities poured through.

Any quality that is part of the parents' motivation as they conceive becomes part of the child's subsequent nature. The child can always change that later, but this is where its first nature comes from. This is the origin of the "first man who is earthy" that the Apostle Paul said must become transformed into "the second man from heaven."

This also is the origin of the historical world.

While we, your spirits, have come from the heaven of the Creator's consciousness, your egos were conceived by our presence in specific regions of the earth. They were born in the ocean shallows of bays now long forgotten, raised on the warm deltas of rivers that would remain yet unnamed for hundreds of thousands of years. They huddled in the caves of the Pyrenees, multiplied on the steppes of Russia, captured fire in the mountains of China and cultivated the once-lush fields of the Sahara. They are the descendants and remnants of those egos who did not understand the original instructions. But, for all their blindness, they are creatures of intelligence, creatures whose mind/brain circuitry has been designed with the capacity to accommodate universal awareness.

Even in the darkest hour of the human drama, we knew that it was only a matter of time. Eventually the age would dawn when they would remember us and invite us back into their lives.

15

Beyond Belief Systems

*H*uman history has been a partial and incomplete description of the universe, a humanly described fiction based on the observation of those without perception of the finer, subtler energy frequencies where the perfect designs of the spirit world are evident. The level of perception that human history has denied is the level upon which we of the Bird Tribes live. It is the level of perception from which we create. It is the frequency level that is your true home, the heaven where simultaneous awareness of your universality and your individuality occurs easily and naturally.

Egos alone, without consciousness of their spirits, are far more limited in their perception than those who merely have no physical sight. Historical egos believe only in themselves and in other egos. They create a world wherein ego values are all that they consider real. There is nothing wrong with ego values, but they are only one side of the universe.

The other side of the universe awaits human discovery and recognition. To 20th century humans, this amounts to nothing less than the discovery of a whole new world. This discovery will transform the face of human experience far more thoroughly than any geographical or technical discovery has ever done in the past.

Historical human egos know they are unfulfilled. They are looking everywhere for that which they lack. Those of you who remain locked into exclusive identity with them put out little grappling hooks in search of your missing pieces, but you continue to grapple at the wrong things. You grapple *externally* in the three-dimensional world, looking for fulfillment outside of yourself. But it is only the spirit of God that can bring fulfillment to an ego, *and the spirit of God is experienced, not outside yourself, but within.*

When you experience God's presence within, you perceive the kingdom of heaven without. The intelligence that the earth loves and craves is present again in your own firsthand experience! You receive internal media far superior to anything external: the channel where forever flow the thoughts of God. And because you hear them, they can flow out from you and be shared with others. As you open to it, moment by eternal moment, a new intelligence, like a whirlwind, swirls up your illusions and restructures your perception. You no longer experience yourself at the mercy of the world around you. The world becomes your paint and canvas. You know yourself to be an artist, capable of recreating all you survey.

There is something more exciting happening here on this earth than any interpretation that ever could be conveyed externally through the medium of an individual human. The reality that we inhabit is a shared field of collective perception that individuals can exist within, but can never "possess." There are many perspectives from which to view reality in this new world. Each individual describes and perceives from different points along the circumference of the great Medicine Wheel of eternal being, but each one does so within the context of certain basic recognitions.

Chief among these recognitions is that all creatures inhabit and live within a single field of shared consciousness, that all are projections of a single Being, and that all of us—angels, humans, animals, vegetables, microbes and minerals—are differentiated aspects of one conscious and coherent whole. This recognition is the cornerstone of the new Sacred Reality, the foundation of the kingdom of heaven.

The second recognition follows from the first: the parts of this single Being, *when they are healthy*, do not complete, but fit together to support one another like the cells and organs of a healthy body. The rich variety of their perspectives, when expressed simply and naturally, complement one another as parts of a whole.

Every angel, every human, every creature, every plant, every microbe within the reality that we recognize knows that, since they each come from a single Being, if they are healthy, they are in harmony with everything else in that Being. Herein lies the source of security.

This is the peace that surpasses all understanding, the inner peace of which your mystics speak, the peace of mind and heart that all true men and women of God have discovered and known. It comes from remaining true to one's own spiritual nature and thereby remaining in creative harmony with every other healthy part of the universal whole.

There have been examples of this awareness, even in your warrior societies. But in a fearful culture, one who is in harmony with God is not necessarily in harmony with the surrounding society. Often the presence of an Awakened One puts pressure on the surrounding disharmonies, bringing them to the surface, flushing them out, like a cold will flush out toxins a body may need to remove. Eventually their presence brings peace, but not necessarily at first. This is why Jesus said, "Yes, I come to bring peace, but first I bring the great Sword of Truth that will bring illusion to the surface where it can be recognized and severed from that which is real."

Before the peace comes to the human world, unhealthy elements have to depart. Yet this does not imply apocalyptic scenarios. Nothing physical has to depart. Not one single creature has to depart. Nothing has to depart but ideas—ideas that have distorted the human tribes of the earth and twisted them into societies of greed, dishonesty and violence. We are here now in human consciousness removing these false ideas. Many of them have already been removed. Before this purification cycle is over, the very last of them will depart.

In cases where people identify with illusion and fail to make the distinction between themselves and what is being removed, by their own choice, they choose to depart during these times along with the illusion they have created. There is never any need for this, but there does come a time when those who would deprive others of their right to biological expression cannot, in rightness, in justice, in harmony with the Sacred Way, be protected any longer. They need to continue their lessons in contained ecological niches where they affect no one but themselves.

For thousands of years we have been protecting those who serve the gods of fear, protecting them from reaping fully what they have sown. Had we not, the warrior tribes would have long since perished. Now we remove that protection. The year is 1987. The time is propitious. Our millennia-long education has at last created optimal conditions for the purification of the human world. During this purification, the warrior tribes will not perish, only their habits. This is necessary. We will not continue with predators here, or with those who cause others to fear, but with those who love one another as they themselves are loved by the Creator of all.

In harmony with the intentions of the Great Spirit whom we serve, we shall protect the rights of all to live and have life except those who persist in the worship of values other than the values of life. These shall harvest the results of those values—experiencing themselves the direct and now unmitigated reflection of what they do to others.

It is we who extend permission for everything that occurs on this world. And just as we extend permission, we refuse it as well when a certain educational saturation point has been reached.

Long before the Mayans made note of it, we set in motion the influences that would one day end the subconscious human condition. It was not difficult to calculate when the point of diminishing returns in the healing of the warrior tribes would be reached.

Five centuries before the dawn of the present era, in the jungles of the Yucatan, we brought to the Awakened Ones of that age a timetable, which they carefully recorded in stone. The Mayan Calendar is recovered now from moss, fern, lizard and leaf. In it are chiseled the dates of the Great Day of Purification, dates that correspond to your years, 1987 to 2011. The winter solstice of the last year of this intensified twenty-five year cycle will see the purification complete, the era of human history brought forever to a close.

We are here now in growing influence to free your race from the dominance of historical ideas: ideas that produce war, ideas that create poverty, hunger and disease. We are slowly, but absolutely removing these illusions from the field of human consciousness.

Because people share the creative power of the Great Spirit, their thoughts are creative. When people conceive of an idea, such as the concept of "survival of the fittest," that idea takes on reality for them. It becomes part of their description of the world. They inhabit that description. They invest their creative power in its illusion.

This is what it means to "believe in a false God."

There are many such fictitious premises that are worshipped as the false gods of the human world.

To the extent that people believe in such illusion, they create living situations that cause them anguish. They live in inner torment. They cannot possibly live by the myriad ever-shifting values required by these "popular idea deities." They cannot reconcile the loving energies of their life-force with the fearful concepts that society would use to rule their lives. They are people divided, people in pain.

Disease does not cause pain; disease is the result of pain, the result of spiritual agony. Disease is a subconscious strategy created by those who are spiritually suffering; it is designed to free them from the false gods—conflicting values and ideas—that hold them in their grip. But there is no need for such an inferior, subconscious approach. There is an easier way to free oneself from counterproductivity, from illusion, from the tangle of history's lies.

On the currents of eternal love, we bring an invitation to a reality far beyond the passing deities of the human mind. We remove erroneous ideas from the field of human consciousness, not by attacking them one by one, but by helping people to experience the truth that alone can replace them. That truth is metaconceptual; it is beyond the realm of conceptualization. Ideas can reflect portions of it, but they can never contain the magnificence of its living wholeness.

Healthy intelligence never tries to hold on to the truth. Can a fish hold the ocean? Can a bird hold the sky? Healthy intelligence knows itself as a center of creative description, living within an infinite ocean of truth that stretches forever in all directions. Healthy human identity does not rest in a conception of the mind; it rests always in spirit.

The development of a more accurate understanding of the universe is a necessary *aspect* of your awakening process, but it is not by any means the *essence* of it. Your challenge in this age is not to formulate new and better descriptions of reality. Your challenge is to shift your very sense of self back to the primal, undifferentiated unity of Spirit. That is your essence. That is God. Out of that *eternal sense of being in love* creative perception flows: a stream of awareness, teeming with the idea life of God. But the ideas that ride upon the eternal outpourings of that great love are there to serve expression, not to dominate it. They are your servants in the dimensional world, designed to assist you in the embodiment of your love. They are not masters who define you.

Your belief systems have been your historical masters, feudal lords from whom you must now seek liberation.

Make the new and vital distinction between *your under-standing* and the terms in which your understanding is expressed.

Your belief systems are not your understanding. At best they are but vehicles of understanding, vehicles of expression. There are times when a vehicle is convenient, but there are many more times when the vehicle is left behind. You may travel occasionally in an automobile, but you do not stay shut up inside it for decades at a time.

If you need to travel through certain dimensional relationships, you may find it useful on occasion to climb inside the structures of your understanding and use them for a while like a vehicle. You will always be upgrading these structures. You create new models every time you understand the universe better. Structures of understanding are necessary. You need them to communicate, to navigate, to function as an incarnate human being.

However, *you* are not your present structures of understanding. You inhabit a multidimensional reality far beyond any structure of thought that could ever be created by your human mind. Know that you are not your thoughts, you are not your ideas, you are not your descriptions of the world around you. You are a being of pure consciousness who has chosen to manifest on the physical plane. You have incarnated to bring out, develop and enjoy the beauty of the created realms.

In any given moment of time, the forms through which you understand yourself and the universe naturally reflect different observations, different valleys, different geographies, different landscapes and societies. You will always have these forms of understanding the world in which you live. They can be good things, like your ego. But in a healthy state they are things that you consciously animate when and how you choose. In healthy function, you are never trapped within your concepts.

Whenever you choose, you walk away from your maps and models for times of being, times of insight, times of peaceful attunement with All That Is. On these occasions you leave behind your beliefs and ideas, no matter what culture or society or geography you are living in. You leave behind all the forms of your understanding, no matter how accurate you believe those forms to be. You leave them all behind, mere foam on the surface. You relax into a deeper knowing. Your identity expands beyond your individuality. You experience yourself in perfect unity with the eternal Spirit at the source of your life. You rest in that knowledge, in that oneness, in the infinite ocean of eternity's peace.

If you have left a biological body behind, when your times of attunement and meditation are over, you return to it and to the forms of understanding it creates and entertains. And you entertain all these forms of understanding as guests in the lodge of your being.

You experience your individuality as a healthy leaf. You feel the sap flowing from the vine of your innermost self. You feel your life force coming from the central core of your own eternal love. You are the leaves and branches in time and in space, but you are the vine in spirit and in essence. Should ever your individuality perceive a problem, relaxation into universality provides the answer.

Frequent communion with the great eternal Being at the center of all life keeps you clear, pure, connected. After such times of communion, you go back to your individuality, either knowing that it is time to release this one of your many forms, or you return to your individuality healthy, alive, vibrant, full of life for the season in which you thrive. You choose if your forms of expression shall live or die. And those forms you choose to animate, you empower with the fullness of your life.

16

To Walk Upon
Troubled Waters

We of the winged tribes arrive in this age, *not just to materialize, but to incarnate*. We are coming back at this time on a wave of light, a pulsation of new intensity. We have the ability to materialize through drawing atoms and molecules into our light fields, but it is not for this that we are here. We are looking for biological incarnations in the specific human beings whose present bodies have grown from fetuses that have unfolded along the vibrational patterns of our light.

It is we who designed your human bodies, but you have grown up in a civilization that denies both our presence and our reality. And so we breathe in you again and again, each breath fuller than the breath before, each pulsation brighter, each new communication clearer, penetrating further into your understanding.

177

Let us wake up in you, we who have formed you in the womb, we who have called you from the tomb of sleeping matter. You, whose physical circuitry mirrors our being, welcome us into your consciousness.

When you experience our spirit presence, you know yourself in the image that God holds for you, the image of perfection in which God has created you. God's living image of you reflects God. It does not rest frozen like a photograph in a frame. The perfect images in which God holds each human being are alive and changing, flowing, fluid images. One day a servant, the next a lover, the next a sister, a brother, a teacher or a friend.

We understand God's images for humanity, for that is what we are! We are individualized, living images of God, designed to draw more than just your physical bodies into our energy patterns; we are designed to blend with your consciousness, with your understanding, and with the field of your emotional expression. We are here in the atmosphere of your consciousness to draw you into clear and accurate perception, that together in cooperative harmony we might commence a new cycle of creation.

We come to the earth to blend with you in the communion that will give birth to the Creator in time. We work with whatever forms of understanding you allow us to animate. Our love has called up out of the earth the bodies of your race. Feel that love and know your spirit. We offer you more than words; we offer you ourselves.

O children, we have lived and died in so many of you. And we live in you again, again and again. Your fearful human generations are like waves of cold sea water, splashing on our warm eternal shores. You can keep being afraid and receding with each cold wave to incarnate again and yet again. Or you can begin to watch the waves, the waves of your generation's fears, the waves of your society's illusions, the waves of emotional control, and realize that *you are not the waves.*

The humanly defined world is constantly changing. It always has its momentary drama, a wave of emotional turmoil that seems to have the ability to put fear into the hearts of more than ever before. But no matter how wild and turbulent the waves of your culture's gods become during these last days before such illusions vanish, be not the waves of fear, nor influenced by them. The birds have always floated on the surface of such waves. Even in the worst of tempests and storms—storms that have crashed armadas, sunk galleons and seen angry warriors perish beneath their fury—ever the birds have floated undisturbed, above the wreckage, above the turbulence, above the waves, gentle, serene.

But times need not be troublesome for you humans of this closing age. For wherever there is just one that does not succumb to the emotional undercurrents of fear that would herd you like sheep into some collective folly, there radiates an influence of peace, stability, healing and blessing. God will amplify the influence of that one and blend it with the influence of others who are incarnate and doing the same. Through these will pour the peace. And no sea will rise against them.

We are here to calm the troubled waters of collective emotional storms, to walk upon them in our understanding, to bring the ways of love to a human world tossing still in a troubled sea of illusion. We bring the torch of peace that comes from trust in God. We bring the torch of wisdom that resolves human differences through commerce, communication, forgiveness, honesty.

Loving our enemies, we have no enemies nor anything to fear. Our torch lights beckon from above the waves of human turbulence. The consciousness of the Eternal One shines forth from each light we hold. Those who struggle and clutch at wreckage in the turbulence below cannot harm us or our kind. They can only become us in time. For it is from them, from such raw material, that the egos of our universal species are drawn.

Do not subscribe to the passing illusions of this age. Do not blindly follow the archaic warrior programming that leaves you unconscious, an automaton, a puppet controlled by those who have for centuries made it their business to map and chart the currents of fear. Realize your choices. Should you feel a habit of violent reaction, whether in thought, word or deed, realize that it is only yourself with whom you strive. Do not contend so anxiously with yourself. There is only one of you here, shadowboxing on the wall. Defend yourself from this alone: the programming of unconscious fears.

Every time a human being reacts violently toward another being instead of communicating, that person is hurting him or herself. The essence of intelligence is simply this: it is always to your advantage to cooperate rather than to retaliate.

These are not "others" that you perceive as foes, but those of your own household, your family. Let them become your friends and partners in the healing and educational work of these times. Do not take them or yourselves so seriously. With lightness in your heart, rise above the waves of emotional concern that would draw you beneath their surface. Feel the love that is restructuring every human illusion.

We come in this age to calm the waters of human emotional turbulence. We have allowed them to swirl this past while to help the egos learn. But the Creator has appointed an end to history. And we are the means of the calming. The water is calmed wherever we stand. We walk on the water singing. Pure water: a human heart in love. Clear water: the eyes of God all-seeing.

Standing above the swirling seas of illusion, you see the radiant light of ten thousand other beings such as yourself. Light beings. Winged Ones. The Ongwhehonwhe, returning. Your love for these sisters and brothers of old bursts from your heart. Rays of brilliant light shoot between you. Ten thousand bridges of light connect your hearts. And beneath each luminous bridge, the light falls laughing on the swirling seas. These waters, too, are calmed, the waters beneath the bridges.

Up out of the angry seas, others rise to join you. They leave behind their petty claims upon this, that or the other moment. Into one larger moment they merge with you, into one moment magnificent. They feel the network of light that engulfs the planet. They become a part of it. They awaken into the energy of the children of light. By your side, they experience the teamwork and cooperation of the real peoples. They see. You see.

Something is touching down upon this earth, something from the stars. Something is landing upon the still waters of hearts that trust in God, speaking to the hearts of people who love.

Each splashing wave of time brings another moment, another doorway, another opening, inviting the awakening children of the earth to feel the flutter of spirit in their consciousness, to feel the brush of angel wings, to allow the waters of their hearts to be still and to live each moment with greater love, honesty, caring and compassion than ever before. So it is that the balance of the Sacred Hoop is restored. Moment by moment, *you* become more incarnate.

17

Medicine Wheel of Eternal Being

When you examine the nature of that pure essence of beauty behind the perfection of both masculine and feminine, you find God. Because at the source, at the core, at the heart of all that is feminine is God. And at the heart, at the core, at the source of all that is masculine is God. The same God, the One, the Great Spirit. Such is the nature of God.

God is indescribably beautiful, indescribably wonderful, indescribably precious. Powerful. Gentle. Real. God is these qualities to such a perfect extent that God's only challenge is to pursue description of such indescribable perfection.

So the Great Spirit picks a place in the infinite ocean of space, draws a line. And calls it time.

Then, making the first distinction, God says, "This line separates everything that happened before I drew this line from everything that happened after I drew this line."

And God takes all the qualities and potentialities of eternal being and places their centers on either one side of the line or the other. These qualities continue to pulsate on both sides of the great distinction. Their radiations permeate all that is. Nowhere is without them. But the centers from which they radiate are assigned places on either side of the great line.

On one side of the line, God designs structures to identify specific locations in the sea of infinite being: atoms, each an intention, a consciously created system for cultivating and regulating its particular field of space. On this side of the line that God calls "time," bleeds through all that is feminine, material, actualized structure. Here lies all that is created, everything that exists, the stars and the planets. This is the Tonal, the body of the great Mother universe. Home of every star and every world.

The other side of the line is the Nagual. Here remains all that will be created. Here are the patterns of new creation and the energy that will draw the new creation out of the infinite reservoir of eternal potential, from the designing intelligence of the One who is ever becoming revealed.

Yet the line that God has drawn to define the sea of space is not a rigid, linear boundary; it is a living reality, spiraling like a thread in and out of the created realms. Around and around the circumference of every namable thing, it coils and spins, weaving a fine luminous cloth, the filmiest of screens, an invisible membrane. At each point where it touches the dimensional universe, incredible creation occurs: stars are created, galaxies congeal, white holes appear, black holes, quasars, arching starbows. In a multitude of awesome ways, energy pulses through from the masculine side of the great spirit divide into the body of the Mother universe.

In this way the purpose and the passion of the Creator are served: the entire universe becomes a joyous journey through love, an adventure, a challenge to describe the infinite wealth of invisible beauty, to give it texture, tone and light; to make it visible.

Creatures of beauty come into form; not just isolated creatures of beauty, but creatures of beauty each living in a larger landscape of beauty. And not just landscapes alone!

Landscapes resting within picture planets, framed in spiraling systems of stars that are themselves set like living diamonds in a galaxy floating among a hundred billion more, giving each creature a background, each picture a frame, and all living art a context that accentuates what is most beautiful. Division into stellar and planetary polarities is the physical mechanism of this awesome creation, the manner in which the infinite potential of God in all its splendor is revealed.

Stellar and planetary polarities are not "opposites." They are complementary ends of a single vibrational continuum, ends of a unified spectrum that begin, back to back, at a common point on the great Medicine Wheel and then move apart, each to complete nearly—but not quite—the full 180 degrees of their respective halves of the whole.

Just short of meeting, they stop to face each other, 359 degrees of the Wheel complete. They pause then, allowing the charge between them to build and grow until there comes the spark of lightning that completes the full 360 degree circle of the Medicine Wheel in the illumination that here clothes you in biological form.

There is design in both matter and energy, just as there is design encoded in the genetic structure of ovum and sperm. The design that enters the earth's atmosphere on the rays of light from this nearest star calls to the potential of this material world; the stellar stream touches her deepest ocean dreams and sings lovingly to the longing in her soil. The information contained in the elemental forces of the earth mingles with the spirit information of the solar wind, as feminine and masculine polarities of the Eternal One play together and interact in the biosphere of Etenoha.

Sunlight impregnates the earth. Solar sperm and ocean ovum mingle. The sun teaches the earth his song, and the earth teaches the sun hers. The sun tries to sing her song and laughs. And the earth tries to sing a sun song and creates blue sky, white clouds and a world of scurrying creatures below. Our spirit information blends with the material information of soil, water, wind and sky. The gestation of the terrestrial child is catalyzed. Creatures rise up on the spiraling winds of our vibrating love. Feeling our song, they live, breathe and dance out the patterns of our interaction.

Spinning in starlight, the earth learns a new image of herself. She releases her rock formations to flow into waving forests, leaping deer, grazing buffalo, sleeping lizards in the sun. Particles of her matter are given wing and flutter and sing on the winds of solar spirit. The earth releases who she is to become the life dancing upon her surface, flowing, ever flowing into the blossoming patterns of her potential.

From the realms of higher vibration where forms are subtle and have no material density, the angels of the Bird Tribes descend. We seek to discover how we might most effectively and creatively clothe ourselves—and thereby the Creator's intentions—in the matter of this world. We watch our leaping flames of fire lifting the rocks of her surface, rising beneath the density of her structure, growing, as the seasons pass and our love deepens, into forests waving in the wind, teeming children, animations of starfire, clothed in the beautiful, feminine, crystalline latticework of the material plane.

Dialogue between masculine and feminine aspects of the Great Spirit has been going on in this universe for twenty billion years. The creation that has appeared as a result of this dialogue has appeared on each vibrational frequency of manifestation in turn, with degrees of specificity and precision on each frequency directly related to the quality and clarity of the exchange among the embodied representatives of male god and female god who there come into being. Once they come into form, embodied representatives of God become the primary means of all future creation on the frequencies where they appear. They are the specified, intelligent and co-creative paint brushes of the universal artist: living, participating art.

It was long prophesied, even before the foundations of this material world upon which you now stand, that a day would come when the loving interaction of a planet and a star would bring into being autonomous, self-replicating, intelligent creatures, biological creatures—male and female, after the Creator's image and likeness. These would become, not only our participants in this continuing process of divine unfoldment, *but also henceforth the leading, cutting-edge in all further physical plane creation.*

They would understand and embody the balance. They would comprehend the nuclear energy of stars as well as the geological formations of crystals in the heart of the mountains. They would understand organic chemistry, nuclear biology, astronomy and all the sciences. They would be incarnate in bodies—physical forms that were beautiful to look at, graceful to behold, living works of art themselves, designed as biological control rooms, mind/brain planning centers for exploring and developing ways and means that the potential of this universe could continue to unfold. They would become the instruments through which all future biological creation would occur.

On the level of this earth, in the late 20th century of the present era, there are biological creatures of female and male human form at either side of this profound interface between the masculine and feminine aspects of God. The same division between eternal unity that has the power to bring stars into existence ten thousand times the size of your sun, that same creative charge exists between the human female and the human male.

In a healthy state, the power of this creative energy is immense. Consciously channeled, it can be used to create beauty of an unprecedented order, of a nature surpassing all that came before, making visible the invisible beauty of God. Never before on the material plane has dialogue between divine masculine and divine feminine had the potential to be as clear, as concise and as conscious as it can be between the men and women of your human family.

When male and female meet on the Medicine Wheel, they each carry a distinct, creative charge. Though their roots weave together and spring from the same source in the same Great Spirit, they incarnate on earth to face one another with distinct sexual charges, profoundly attracted, with immense capacity to create through the universal power of the love that can flow between them. Each woman swims in an energy current, a tradition as distinct from the traditions of men as the very first creative division of God. Though you men and women have lived together in the same lodges and felt beneath your feet the earth of the same villages, though you have shared the same interpretations and slept together in the same illusions, you are each at distinctly different points on the great Medicine Wheel. Yet as different as these points are, they are profoundly compatible, designed differently to stimulate maximum creativity. They are the very points where God meets God.

To we who have followed this process for twenty billion years, there is nothing more incomprehensible than the concept, which is entertained in the warrior societies, that there could be conflict or opposition between the feminine aspect of God and the masculine aspect of God. Do you realize how severely this concept limits you who accept it?

You will not find your place in this universe or form a basis for understanding her mysteries, until you shed such superstition and *realize that Mother God and Father God are profoundly in love.*

All about this universe that is feminine worships what is masculine. And the sole purpose of all about this universe that is masculine is to serve the feminine through the celebration and animation of the beauty that lies in her heart. Feminine and masculine are balanced in all healthy manifestation, equal partners, lovers, the truest twin friends.

Your ancestral programming is invalid. Primitive notions of conflict between male and female lead to not-so-primitive weapons. Let your human world reflect instead the truth of the great love in which you are every moment suspended. Awaken into a new perception and understanding of reality: reality—not as misinterpreted by the warrior tribes—but as understood by all the angelic races of God throughout eternity. Awaken into an accurate perception of the universe in which you live.

You are the love of the Creator embodied in human form. Through the power of your love, you create. You are here in the service of universal art to create beauty and to enjoy all that has been created. *The evocation of beauty and the description of truth: this is your purpose.* The purpose of life! The very purpose of the universe.

In its own way, each species decides how it can best serve this purpose, once it understands it. The buffalo looks around and notices what is going on in prairie time and prairie space and designs a life to describe truth and to animate beauty as buffalos understand it. Whales describe time and space in their own unique forms of three-dimensional mobility, playing and swimming through seas of sound and light. The caribou embody their understanding of the universe and create caribou beauty. Sea gulls embody their understanding and create sea gull beauty. Angels embody their understanding of this universal purpose and become the means through which the Great Spirit creates more species to enjoy it.

Throughout the galaxies, throughout the stars, all of the angelic races notice this purpose of life and decide how they can best serve. Each *healthy* species looks at the external world, while experiencing the internal unity whole and unbroken, and forms an idea of the wholeness of manifest life. This is its comprehension of the universe. Human beings' *technical* comprehension of the universe has expanded beyond the caribou or the whales. They have explored the microbes and begun to reach out to the stars. Yet your historical rejection of eternal spirit has left you myopic, nearsighted, with no overview or larger vision to help you organize your knowledge into a coherent whole.

You are the instruments through which creation is to become *physically conscious*, the species through whom one day the whole material universe shall become aware. You are God's organs of awareness, awakening, in this turning age, from the sleep of historical subconsciousness. Masculine and feminine are both rooted in the unified divinity whose Spirit now awakens in you, whose identity you now remember to be your own. In spirit you are not separate from these two aspects of God, but have brought them into being from the infinite ocean of your wholeness, to clothe yourself, to know yourself as this family in time. Creation is brought into being through the loving interaction of these, your own inner polarities.

18

A Solar Age of Flowers

*T*he conception of new life requires highly focused, creative intention. It takes attention to detail of such a profoundly precise nature that were the focus dissipated among many cells simultaneously, the miracle of conception could not occur.

Of the human body's one hundred trillion cells, only *TWO CELLS* are initially involved in the conception of a child.

You float in a galaxy of some four hundred billion stars and many more matter worlds. Yet in this galaxy, only *TWO WORLDS* are required for the conception of new life.

Do not be surprised when we, who have roamed among these stars, tell you that you are special, that you are precious, that you are so incredibly loved by the source of all that is, that your one earth and your one sun are conception points for the new life of worlds to come. Do not shun such love or let narrow minds convince you it could not be.

Accept this love. Feel it for yourselves. A gathering together and focusing of energies has been necessary to produce the awesome and incredible miracle that you are.

The attention of the universe is focused on this earth. The Universal Mother herself is conscious in this world. The Father of every star looks through the eyes of this one sun. Your earth has experienced human conception.

Your race is about to awaken as the circuitry of consciousness, the global intelligence system of a planetary newborn that will be half star and half material world.

I will tell you, I who am an angel of the winged tribes, I have been consciously creating beauty for twenty billion years; and I have never seen at any moment in all that time, the potential that I see among you people at the close of the 20th century upon this fair earth. I have seen many spirit beings throughout the stars, intelligent societies on the higher planes of many worlds, but they were not *biologically incarnate.* They had not entered so fully into material dress or come to such intelligent terms with matter.

Nowhere but here have I seen such exquisite, sensitive creatures of biology, such graceful men and women who exhibit your capacity. Do not let those of limited vision cheapen the miracle you are living or diminish your sense of value with their abstractions. The potential you embody is unprecedented.

You are the first of species on material worlds where the necessary balance is appearing to support the new blended consciousness, the consciousness that is at once both spiritual and material. No biological species has before looked up of its own and taken charge of itself as an independent, spiritual being. Nowhere else in this galaxy has the maintenance of this precise temperature zone resulted in a *biological* network of intelligent, global-species circuitry, simultaneously both spirit and matter. Never before did the biological creature choose of itself, did the clay speak to us and say, "I am a woman. I am a man. I am a tribe, a nation, teach me."

O humans of this yet sleeping age, there is more. So much more. It is hard sometimes to place this awareness in your words, for your words are of only one dimension, of only one meaning at a time, and the wonderful, living reality of this universe and the things of which we speak are multileveled, multidimensional, with reasons woven together like many streams and waters weave together in the larger truth of a river valley.

There are many reasons why your world holds the attention of the universe. Your initiative, your capacity, your potential, what you have already demonstrated in your hours of vision and nobility are some of the reasons—but your fall, your rebellion, is yet another. For you see, once the egos of the warrior tribes ran off into the jungle, we knew that we were confronting in the womb, in the very moment of conception, certain fundamental issues that would have to be faced again and again on our future travels throughout the far-flung stars.

We were confronting these issues on a miniature, symbolic level; and if we could get the mathematics worked out with precision on the micro-level of this earth; if we could use the very value systems of the rebellious egos to convince them once and for all that every interest that they held vital would be served more effectively through their voluntary cooperation with the spirits of God; if we could convince them of that in their own egoic terms and if that lesson could be not only learned, but also recorded and imprinted indelibly upon the genetic structure of every cell in every existing and yet to be created organism of this new type, well then, do you see? Do you see? *The suffering of human history would be turned to advantage, for it would never have to be repeated.*

So long as this wondrous universe unfolds, human egos will face the same fundamental issues that you face today. Always they will have the choice to be ruled by what they love or by what they fear. The option to behave with excessive fear and the option to reject the input of their spirits will always be present, for human freedom of will is essential to the proper working of the universal design.

However, if there could be a documented case history of the inevitable sequence of events that occur when an ego or a tribe chooses a god of fear, and if this case history could be recorded in a genetic media presentation, narrated, as it were, by egos who had themselves experienced history, then every ego skirting the edge of a fear-motivated decision could walk into the genetic media library and watch the DNA story of what happened here on your earth, observing what becomes of those who choose fear-centered behavior. They could watch a graphically illustrated story of what happens to those who ignore their natural design and misdirect their energy into cancerous and ultimately self-destructive behavior.

The angels of conception who are guiding universal development seek to minimize suffering, your own suffering and that of races yet to come. So you see, when your race fell, our attention gravitated here. Why allow dozens of worlds to experience the warfare, starvation, suffering and deprivation that is born from the egos' rejection of their spirits, when one world would do? And if that one world were the source world, the point of conception for all similar life forms to come, then *there would, after the lesson was learned, be a built-in genetic immunity to such disease in the future.*

You knew us once, O human ones, for the flicker of a millennium or two, when you first left the seas for inland lakes and river valleys—only yesterday it seems—to dress in human form and forget your spirit dreams. But it was time enough. We saw the sparkle in eyes we recognized. And we knew we could never abandon you.

We were patient when your ego-ruled ancestors ran off into the jungle night. We knew that someday your own intelligence would lead you to seek contact with us once again. We maintained the conditions that surrounded this egg of possibility. We held the earth carefully and assigned to her the brightest spirits. In this age our patience is rewarded. There are those of you who at last look up and see beyond the microscopic interpretations of your egos.

You perceive us. We communicate. We do what we can with the words of your ego-created languages, but it is like speaking with our hands through the movement of their shadows on cave walls, the fire of your interest flickering. Crude and primitive symbolism, these letters on a page, but a beginning. For as you sense the reality behind these words, and follow the direction of our thought, there is movement in your life. You leave the cave of history's deceit and enter the sunlit garden that has ever been your true home.

We are the Pattern Beings, the Winged Ones, the Light Bearers. We are the reality of your perfect spirits, coming now to consciously incarnate in your human fields. During the millennia of your unconsciousness, we obeyed your instructions. We maintained the precise temperature range that would produce for you a child, a species, a mingling of star and planet, an autonomous biological-world, creature-being of a new and blended kind. The requisite human life forms now cover the earth; their physical forms are complete.

A three and one-half billion year creative process culminates in your awakening. The human Information Age explodes in unprecedented comprehension waves. Pulsing comprehension waves. Flowing. Into a Solar Age of Flowers. A Starchild rides the crest of light, and bathed in light, is born in love, rejoicing.

Afterword

The day was over. I entered the twilight interior of the lodge and sat cross-legged in a circle of half-naked people. Fascinated, I watched as glowing red-hot stones were, one by one, brought in silence to our circle's center. Motion slowed. The last of the stones was set in place; the opening of the lodge sealed. I could taste the blackness, the silence, the stillness of human space. Time flowed without time, with no measure of its passage. And suddenly, there was light. Dried celery root was dancing upon the glowing rocks, swirling, sparkling like ten thousand twinkling stars in the great eternal path of the Milky Way. Memories rose from some uncharted realm within me to form patterns in the starlike glow. I felt an unnamed power. Intimate. Familiar.

I had often heard of the sweat lodge ceremony, but this was my first experience in a traditional native American lodge. The dome-shaped structure had been constructed on the slopes of Mount Shasta by a member of the Karuk tribe. Each component of the frame was built with the type of wood specified by long-standing tribal tradition. Covered with colored blankets, the sweat lodge rested beside the ice cold, flowing waters of a spring fed river. The sweat ceremony was offered on this, the first evening of a Starseed gathering on the mountain.

It would take many pages to describe what transpired in that lodge. Even in the silence before the first water was splashed upon the rocks, the words in which I attempted to clothe my understanding of what was happening had already become as elusive as if they were dead leaves, caught up in some restless spirit wind. But as the water was splashed upon the glowing red-hot jewels of the earth and the first burst of steam threw its power to the air, a rushing gust of memory seemed to blow the last of those tattered word leaves far into the depths of night. I found myself suspended in a silence, rich and enduring, a silence within which language did not so much seem to cease, as to be given birth.

When the chanting began, I knew the words, I knew the melody, the rhythm, the inflection. I knew their every meaning. Without thinking, I sang from the depths of my heart in the familiar flowing native American tongue.

Not long into the chant, I remembered—almost casually, as if it were a matter of small importance—that I had spent far more time chanting in sweat lodges than I had ever spent . . . for example . . . inside an automobile.

It may sound strange, but when that realization filled my awareness, it was not as if I remembered the sweat lodges and the chanting. Such deep experience my spirit had not forgotten. It was as if I remembered the automobile—a flickering recent image soon released to the steam and the katydids and the bull frogs and the sound of water rushing over rocks in the moonlight. Eternities passed. Those of us within that lodge had fused into a single being, a single heartbeat.

Time after time, water was thrown upon the glowing stones.

Without warning, flowing out from deep within me, came memory's unbidden explosion. Suddenly there were other stars dancing in a field of midnight—not dried celery root this time, but real stars in a real sky, high above a place I loved more than I could ever say.

I was one of a circle of luminous beings gathered at another meeting. The thoughts that were shared were but part of our purpose. The very arrangement of our circle and its placement against the stars was itself designed to imprint upon us something more; so that later, if a time came when we found ourselves slumbering beneath a blanket of history, the very image of the Circle of the Faithful of the Stars might shake something loose in us and jolt us into wakefulness. My present purpose in drawing attention to this meeting—and sharing this account of how I came to remember it—is to call those who attended it back into mindfulness of the sacred trust that the Creator of all this universe bestowed upon us at that time.

You may remember. The location of our meeting was not upon the physical surface of the earth, but upon the higher frequencies of the spirit world—a location in the earth's upper atmosphere above the 33rd degree of northern latitude. From where we gathered we could see below us the steady rotation of the Northern Hemisphere, the passage of its oceans, islands, continents.

Those of us present at this gathering were not at that time projecting identities upon the earth. None of us was a graduate of the clash of "reality cooperatives" that had manifest as a second world war. Nor, had we known of its occurrence, would any of us have approved of such an extreme approach, for jolting incarnate projections into wakefulness was a strategy alien to our ways of thinking and, in any case, for us it was unnecessary.

We were spirits with successful experience in securing the cooperation of our human bodies, minds, and hearts during times when we clothed ourselves in form. For this very reason, most of us had not walked the earth for some time. (And among most at this particular gathering, it would require calculation to say whether that time was centuries or millennia.) A few of us had helped to guide the human family through its more recent turning points, but we were the minority.

The meeting was called to order by the great angel whose body is the sun, a customary procedure that did not surprise us. Typically however, such meetings were held only for those about to embark on incarnate lives—and there were few among us with that intention. We were alerted to the unusual significance of the occasion, when flickering across the face of the sun we beheld the countenance of the Universal Mother, and indeed, when the thought transfer began, it was she who spoke, she whose very body is the galaxies. Her purpose in addressing us was introductory in nature and proved of value in preparing us for what was to come, for I think without her intermediary guidance, what appeared next might have otherwise triggered so much surprise as to cause, at the very least, a delay in the proceedings: it was none other than the *white light of the Eternal One,* focused in a manner rarely seen in any star.

We lived in the presence of that Great Spirit, it is true. Our lives were lived knowing it as the source from which we drew our all. But to witness so sharp a focus of that presence was as distinct from what we were used to as gentle sunlight is from sunlight focused through a lens upon a single spot.

In the communication that followed, the three manifestations of Being—solar, universal, and eternal—flowed in and out of one another, blending, merging, weaving the colors of their whole. At times it was hard to identify which of the Trinity was the speaker, so unified were their waves and so aligned their purpose. The Creator spoke through these three, giving us specific instruction. I will pass on the words as best as I can recall.

"You are being invited," we were told, "to incarnate upon a world where illusion prevails in the minds and hearts of those whom I have created in my image and likeness. You who accept this invitation will soon find yourselves in the last days of the human species' infancy, in a situation where the currents of fear have ruled their experience of separation. I have numbered the days of this condition and they are soon to end. I have established the external forces that will divide the fear from the love in human consciousness, banishing the one and redeeming the other.

"I am inviting you to be my agents in this fearful culture that you might do what you can from within to complement the changes I am bringing from without. For I know that there will likely be fearful ones whose physical bodies will not survive the separation between love and fear, but I know also that *there is no reason, no good reason, why even so much as a single human being should disincarnate during this cleavage.*

"I have called you here because you have proven in times past that you can, to a high degree, keep your bodies, minds and hearts in harmony with your spirits. I release to you now an invitation to incarnate in these tribes who inhabit the earth. I have not asked this of you until now, for during earlier historical times conditions on earth were not as conducive to the spreading of our awareness. But the awakening in human consciousness has begun. Many are now opening to the living information. Your presence would do much to minimize the possible trauma that could occur with the changing of the age."

I remember distinctly.

While it was being explained that what was happening on earth was like the changing of a guard, suddenly the spinning planet, which until now had seemed unremarkable except for the gentle rocking of her seasons, was sprouting an enormous wound. An ominous gray circular cloud rolled toward us with the new day, obscuring the southern portion of the Japanese archipelago.

As our vision penetrated beneath, we saw that it was supported by a black pillar, a turbulent inferno of awesome destruction. With each roll of the earth we could feel raw human terror broadcast upward, rippling through our open hearts in waves of unbridled fear. Then suddenly, rolling toward us from the west, came another. A towering column of dark red, cloud-like substance surrounded an empty void that seemed as impossible in the midst of such chaos as the chaos itself.

Immediately we slowed our time perception and zoomed in to experience this day. Into a vacuum hole, soil and water and organic life poured, tumbling and boiling, rising into billowing clouds of turbulence that rose almost to touch us.

As far as we could see, all beneath us was inhabited by swirling, sucking whirlpools and shooting balls of random fire, the likes of which I have seen only upon the surface of stars. I will not elaborate.

Into those moments we brought eternity. In them, we are still piecing together what justice requires. It is enough to say that on that day, we experienced hell. Yet even as we did, we knew what our response to the Great Spirit's invitation would be. Yes, we would incarnate. Yes, the time had come.

The purpose of our incarnations, it was explained, was to begin disrupting the false, underlying premises of the historical state, educating people and gently introducing the truth.

"Your role," our narrator continued, "is to prepare people for the time when the lies that give birth to all fear-centered thinking will be banished forever from human consciousness. Though it may sound strange to you, if human populations are not prepared, it will be much more difficult for them to survive the cleavage, so entwined are the thoughts of fear, not only in their thinking, but even in their very sense of self. We must help them relax with their identities and stop defining themselves as vulnerable entities."

We were told that to effectively achieve this objective, we were, of necessity, going to have to incarnate, not in protected locations as had been our wont in ancient times (having as little to do with the fearful as possible), but into the most fearful of the planet's ruling families. This would ensure our strategic placement among those elements of humanity who were in fact expected to prove most resistant to the powers of education.

The Creator was careful in making it clear to us that we were foregoing our traditional family bonds in order to incarnate in this way and that we were likely, upon awakening, to find ourselves in unsupportive and vibrationally alien surroundings. It was emphasized that we did not have to accept this assignment, that our role in this was purely voluntary. Even if we accepted, there was no assurance that we would succeed in entirely eliminating the trauma of the transition, but there would be a greater chance that more people would come through the moment of final cleavage with their senses intact. To alert us to this opportunity, our circle had been called to gather in the starlit skies above the 33rd parallel.

We were cautioned about the dangers we were likely to encounter in a world still very much under the control of fearful organizations.

"The real power," God explained, "is in the love that is strengthening each day during this coming cycle. Our strategy now is to allow the organizations of fear to continue to dominate in appearance while we subtly drain away the source of their power—the fear of the people. This will make their final banishment less traumatic for the human family as a whole, for by then, the numbers of those who believe in them will have greatly diminished."

We were reminded then of the many layers of fearful logic that we ourselves had shaken off during previous walks upon the earth. We were asked to take who we were and what we knew and to incarnate among the nations of the 20th century. Our mission was to retain our innocence while interfacing with every known form of human organization. We were assured that the momentum of the age was with us.

"You do this not for your species alone," we were told, "but for yourselves and for all whom your race will influence in the ages to come. Though many of you are incarnating among aspects of humankind long kept separate from your own realms of experience, you are nevertheless bound to the warrior tribes by ties that were forged by your own thoughts in eons past, ties that you have only temporarily been able to lay aside—but this you have known from the beginning."

The Circle of the Faithful of the Stars was invited to incarnate *en masse* as a single global generation.

So that the transition might be as easy as possible for those upon the earth, we agreed. Our purpose was, and is, to prepare the way for the Creator's awakening and to dissolve in the great love that we have for this blessed planet, the structures of fear that bind human consciousness.

We would enter human forms, some to incarnate in the next wave of children and others in their children's children. We would do what we could to help guide the world peacefully toward that moment in time when the Creator would awaken inside the human family and began to direct it as a single organism.

"The danger for you who are to incarnate into the fear dynasties of these times," it was pointed out, "is that these structures and the energies that they embody are so very, very subconscious that it is possible that as you are born and raised among them, you will forget your very purpose for incarnating. As your generation reaches maturity, I will give you some special assistance to help you remember."

Implanted then into the structure of our emotional bodies were nonverbal, timed-released *melodies* that would help us to awaken even when conscious understanding was far from present. A number of us were musically commissioned and given the role of introducing the trigger melodies.

These melodies began being delivered in the 1960s through songs that—though often rather nonsensical on the surface—pulsed with powerful emotional messages of love, hope and joy. Through music many of the sleeping began to stir. The first rays of dawn were heard on the radio before many even realized that there was a sky to look to.

Yet our awakening is not complete until what has found form emotionally also finds form conceptually. What we know in our hearts requires the complement of conscious understanding to effectively approach the healing that lies before us.

It is for this reason that I share these thoughts and memories. Our generation is now in the position to steward the earth and her resources. Will we foolishly imitate the ways of elder merchants and statesmen who are themselves only aping the ways of feudal lords now thankfully removed from power? Or will we accept the commission to remake our world in the image of love?

Now that I have surfaced in this age, so long after the feudal lords have been removed, it surprises me to find the momentum of their ways still lingering on and attracting fresh human attention. Surely these lords must have commanded the allegiance of the earth's people well. It seems that humans have been so long trained for subservience that they now feel insecure under their own initiative! They lapse into following the shadows of these feudal ways even when the shadows are in such short supply they must create new ones just to continue their habit.

The challenge before us is to reawaken ourselves and then to join with others to show through our example the beauty and power of the new way. We can leave behind the twilight stage that gave room for sleepy imagination to postulate that the awakening revolved around some guru or human organization, no matter how beautiful or enlightened. We are called to organize in this time, not around leaders, ideologies or belief systems, but around love: love for God, love for one another, and love for our sacred world.

Those of you who were there with me, remember our circle. It is time to incarnate, to get on with the healing and educational work for which we have come.

We have been seeded far and wide across the continents of the world. We listen to the same music as we drive to work in Hong Kong, London, Tokyo and New York. We cherish the same vision of peace and disarmament as we watch children play in the parks of Cape Town, Paris, Singapore and Rome. We dream the same dreams as evenings slide across Adelaide, Tel Aviv, Rio and Leningrad. We witness the same events from Mexico City, Los Angeles, Toronto and Berlin—even as they are occurring, broadcast live, we watch them. Together. We open our eyes in the deserts, in the mountains, in the forests, in the jungles and cities of every nation.

We are today's people. There is no one today to write our script but ourselves. *History repeats itself no more than any other bad habit.* And no habit, however deep or ingrained, will stand a chance of re-election where consciousness, honesty and willing determination characterize human affairs.

While our generation volunteered to facilitate the transition before us, the task is not exclusively our own. Many have helped. Many are helping. And all are invited to contribute their talents and ingenuity. There is no conflict of interest. There is simply new insight and a sense that it is time for the ages to progress.

Those of us in the United States are awakening within one of the most technologically advanced societies on this planet. With a new consciousness directing the tools and resources of our nation, we have immense capacity to benefit the entire human family. The government of our country can do no more than reflect the values, the attitudes and principles by which we live our lives.

What we see in the mirror, we see in the media. The love in our hearts—or the lack of it—underlines every congressional decision. We cannot overestimate the blessings that we might bring to the inhabitants of the less materially developed nations when our hearts are truly open.

We should not be discouraged by the seemingly slow pace of attitudinal change within our institutions. The progression of awareness that precedes awakening is subtle and not always visible on the surface. It may take some time before out thoughts are reflected in the older and more entrenched human organizations. But I do not see our government as being among these.

For all its conspicuous failings, the United States system of government remains one of the most resilient and flexible of all human institutions. I foresee a day when it will be as responsive to the rivers, mountains, forests and plains of North America—and to the people who love them—as it is today to the world's merchants.

With an awakened—*and truly American*—consciousness, U.S. policy would not continue to reflect a European past. It would certainly not kowtow to short-term commercial and financial interests whose advocates, spiritually speaking, have not yet noticed the "New" in front of York, Hampshire or Jersey—to say nothing of the "New" before the Order of the Ages.

As our generation enters into the corridors of power, it is not too much to hope that the government, which our ancestors created "of the people, by the people and for the people," will become one of our means of expression in the world. It is something to expect.

With gratitude for all those who have contributed positively to our society, we consider the challenges, the blessings and the opportunities that come with our stewardship. Excluding no one, we ask all to join us in taking a new look at how we can work with the tools and resources of our land.

When thoughts reflect the creativity of compassionate hearts, even armed services can play meaningful roles in the mechanism of transition. For example, if approached sensitively, with cooperation from all who might be touched by such an action, could there be a role for the U.S. Army in the transportation and distribution of food within and to the twenty-one African nations where our sisters and brothers are starving?

Is there a way we can use our American influence to minimize violence in the world and to *truly* promote freedom? Is there a way to direct our nation's power independent of the financial interests that are attempting to "use" America in the last half of the 20th century like they "used" Britain in the 19th? Is there a way to put some native American sensibility behind our existing technological infrastructure so it can serve to assist humankind's awakening into the new reality?

We can draw strength from many things in America: from the sheer numbers of conscious beings who are interfacing daily with the eroding order, to the knowledge that for every Martin Luther King Jr. or Robert Kennedy whose expression has ever in our land been cut short, a thousand more have bounced back to resume the work more determined, more alert and better informed. It is understandable that our visionaries might occasionally draw to them animosity and remnants of fear, but this in no way diminishes what truly matters. They have helped to draw the human world another step closer toward the light. If a few such ones pass along the wayside, martyrs it would seem, for a vision not yet full, still, the vision continues on. It continues to grow in strength and power and clarity until the very breast of humankind explodes in one great expression that is simultaneously both a longing and a rejoicing.

It is from here, from within our hearts, that the power emerges to heal and transform. "We Are The World," "Live Aid," "Hands Across America." Benefits by performers, athletes and musicians untold. These and a rapidly proliferating number of non-governmental relief and service organizations are but a few beginning expressions of an emerging American Spirit that is very much in tune with the Mayan calendar—and no stranger to the Christian one.

There is an influence radiating from the American continents, warming the seeds of human potential like sunlight warming a garden in spring, an indigenous consciousness that touches all who inhabit these lands, supporting all our steps toward freedom and love.

Native American consciousness swings on the swings of our playgrounds. It skates through the parks of our cities. It runs deep beneath the rhythms of our music. It permeates the very heart of our sports, both amateur and professional. It speaks of a sacred relationship with the *spirit of the land*, a regard for her indigenous wisdom, an attitude of reverence for this earth that provides our nourishment, supports our families and upholds the necessary structures of our nation.

The consciousness of this land that we call North America finds in us a new and more emphatic expression— and it needs room, space to grow and expand. Not geographical territory, but territory of the soul.

The new frontier is consciousness.

This blessed world, this third planet from the star we call the sun, needs you and me to become all that she can be. We have an opportunity to bring to her a great gift, a gift for which she has been longing since our race first appeared among her dreams. The gift of ourselves. Awakened. Whole. Incarnate. There has been enough deforestation, enough strip mining.

It is spring. The passes are clear. It is time to cross the boundaries that would arbitrarily circumscribe our potential. This land is your land. It is my land. But just as she belongs to us, we belong to her. The Sacred Hoop can no longer be denied. The circle requires completion. From California to the New York islands. From the Yukon to the Rio Grande.

As Sacagawea once guided Lewis and Clark to the shores of the Pacific, an aboriginal consciousness is awakening to guide our technology through the economic and political labyrinth of the next quarter century. Let us honor it. And let our incarnations begin. We are the Bird Tribes, the Ongwhehonwhe, the people of reality.

Epilogue

When the sweat lodge ceremony that first brought me recollection of the Circle of the Faithful of the Stars was over, I found myself in a state of clear, nonverbal awareness. I was the last to leave the lodge. I looked deeply into the eyes of the man who had so beautifully and so unobtrusively guided us through the ancient native American sweat ceremony. Without thinking I said, "Arigatou gozaimasu."

It was only later, as I lay on my back in the ice cold mountain stream, appreciating the moonlight through the overhanging trees, that I realized I had thanked him and he had replied—in perfect Japanese.

Glossary

Adadarhoh—chief of the Onondaga and a primary figure in the founding of the Iroquois Confederacy. Also Tadodaho.

Ayawentha—a form of the name Hiawatha. See Hiawatha.

Aztec—Nahuatl-speaking people who in the fifteenth and early sixteenth centuries ruled a large empire in what is now central and southern Mexico.

Bioregion—geographical area defined by its geological and biological features, such as Appalachia, the Ozarks, the Sahara, the Rocky Mountains; a natural way of distinguishing areas of the earth.

Bird Tribes—spiritual guardians of the earth. Angelic entities. Human spirit (or higher) selves. Beings through whom the Great Spirit creates on the physical plane. Previous incarnates on this planet. Original creators of life.

Cayuga—[Ka':yoo ga?] one of the five founding tribes of the Iroquois Confederacy, also known as the People of the Swamps.

Champlain, Samuel de—French explorer. Founder of Quebec (1608), and consolidator of the French colonies in North America.

Children of the Stars—see Bird Tribes.

Circle of the Faithful of the Stars—a specific gathering of spirit beings in 1945 to confer prior to human incarnations.

Columba—born in County Donegal, Ireland, in 521; credited with the main role of converting Scotland to Christianity; erected a church and a monastery on the island of Iona.

Creator—see Great Spirit.

Dakota—Siouan-speaking tribe of the North American Plains. Also states of the U.S., North Dakota and South Dakota.

Deganawida—[Dega:nɔ̃wi:da?] founder of the Iroquois Confederacy, also known as The Great Peacemaker. *Note: among contemporary Iroquois it is considered improper to speak the name Deganawida except by the designated individual in a sacred ceremony or recitation at one of the Six Nation's Longhouses.*

DNA—deoxyribonucleic acid, the master reproducing molecule of cell nuclei that acts as the carrier of genetic information as it weaves in and out of generations.

Earth Mother—Spirit of the Earth in her feminine aspect. Etenoha.

Ego—subjective body-associated sense of self, having to do with one's past experiences in a given biological form or succession of such forms. The steward of material plane fears, designed to care for the physical body. In a healthy state, the ego is a secondary component of human identity. Second Self.

Eightfold Path—the steps defined by Gautama Buddha that lead to the state of nirvana—freedom from bondage to the material world.

Etenoha—[It'ti'nŏ̃eh] our Mother, the Earth.

Eternal One—see Great Spirit.

Fall—fall of human consciousness from God-centeredness to self-centeredness, an event initiating the beginning of human history that occurred when human egos rejected their eternal spirits. The separation from God that causes death and war, the breakdown of communication between spirit and ego. The egos' rebellion.

Fire Bird—a specific tribe of angels within the Bird Tribes, also (less common usage) one of the many names for God. See Great Spirit.

First Self—the primary sense of self for all healthy conscious beings. The Eternal One, God. See Great Spirit.

Five Nations—five founding tribes of the Iroquois Confederacy: Mohawk, Oneida, Onondaga, Cayuga, Seneca.

Francis of Assisi—born in Assisi, Italy, around 1181, he founded a religious order dedicated to following the teachings and example of Christ; he is (with Catherine of Siena) the principal patron saint of Italy.

Gaha—[Gæ:haʔ] spirit of the wind.

Galilee—hilly region of northern Palestine.

Ganges—a river of the plains of northern India, considered sacred by people of the Hindu faith and others.

Ganienkehaga—[Gañyeʔge:o:nɔ̃ʔ] another name for Mohawk, see Mohawk.

Gayaneshakgowa—[Ga ya nes hæ'go:wa:ʔ] the Great Law of Peace.

God—see Great Spirit.

Goddess—God in her feminine aspect, female deity. See Great Spirit.

Golgotha—hill of Calvary where Jesus was crucified, near Jerusalem.

Great Day of Purification—period during which the earth is purified, fear and love separated in human consciousness. Mayan: the period of time from 1987 to 2011. Christian: the Last Judgment. Final cycle in the healing of humankind, culminating with the Earth's awakening as a single being and the Creator's awakening in human consciousness.

Great Distinction—living membrane that defines the boundaries of the universe, divides existing structure from future structure, separates the Tonal and the Nagual.

Great Lakes—chain of lakes in east central North America: Superior, Michigan, Huron, Erie and Ontario—a boundary presently between Canada and the United States.

Great Line—living, spiraling thread that weaves the membrane of the Great Distinction. See Great Distinction.

Great Peace—a state of peace in all of North America east of the Mississippi River brought about through the Iroquois Confederacy. The Great Peace lasted over a period of time that included the eleventh, twelfth, thirteenth and parts of the fourteenth centuries (however, many historians misplace the Great Peace at a later date).

Great Peacemaker—see Deganawida.

Great Spirit—the single Being whose unfoldment has become this universe, the Source of all Life. Also known by the following alternate and sometimes interchangeable titles, depending on which aspect is being emphasized:

Mother God, Father God, Wakan Tanka, Eternal One, Mother of Old Ones, Goddess, Great One, Holy Spirit, Grandfather of All, Above One, Eternal Being, Spirit of Truth, Great Mother, Universal Mother, Fire Bird, Thunderbird, Source of Life.

Great White Roots of Peace—symbolic representation of the principles that produce peace and unity. The foundation of peace. The Truth. Right Livelihood.

Hiawatha—[Hayō:wẽ tʰaʔ] an Onondaga companion of Deganawida, a central figure in the founding of the Iroquois League. Alternate spelling: Ayawentha.

Hokseda—[Hokʔsaʔdaʔ] higher self, spirit self, see Bird Tribes. Literal: he has this person as his child, or this is his child.

Huron—Iroquoian-speaking North American tribe from the forests along the St. Lawrence River.

Inca—South American tribe who at the time of the Spanish conquest in 1532, ruled an empire that extended along the Pacific coast and Andean highlands from the northern border of modern Equador to Rio Maule in central Chile.

Incarnation—act of a spirit being dressing in physical form through growing a biological body that reflects the spirit body, act of entering into that body spiritually, psychologically, spatially and emotionally.

Incarnational process—the process through which spirit beings consciously enter biological forms, the process of awakening in which the hokseda or higher self stabilizes in a human being.

Industrial Revolution—process of change from agricultural economy to one dominated by machine manufacture, began in England in the eighteenth century and from there spread to other parts of the world.

Innocent III—Pope from 1198 to 1216, convened the fourth Lateran Council of 1215, gave the official approval of the Catholic Church to the proposed work of St. Francis of Assisi.

Iona—a small island off the west coast of Scotland long considered sacred. During the Middle Ages, it was a favored burial place for the kings of Scotland, Ireland, Wales, Norway, Denmark and England.

Iroquois—a group of North American tribes inhabiting upper New York State and surrounding areas.

Iroquois Confederacy—see Iroquois League.

Iroquois League—political and spiritual union of five Iroquois nations: Mohawk, Oneida, Onondaga, Cayuga, and Seneca. After 1722, in a later and less influential revival of the union, the Tuscarora became the sixth nation to join the League, accounting for the modern reference to the "Six Nation League." Also called the League of Iroquois Nations and the Iroquois Confederacy.

Island of the Turtle—North America.

Jefferson, Thomas—third president of the United States from 1801 to 1809, principal author of the Declaration of Independence.

Karuk—native American tribe of the western coast of the United States.

Keepers of the Light Circles—tribes that keep the light of truth in expression. People who are true to the Bird Tribes. Also used to define distinct groups within the Bird Tribes.

Killarney—lake district and town in County Kerry, Ireland.

Kingdom of Heaven—realm of divine ideation, a substructure of pure form in essence. A condition of wholeness in which God and humankind are one. The healthy state. In the Christian paradigm, the abode of God and the angels and the souls of those who are granted salvation.

Lake of the Michigan—one of the Great Lakes between Michigan and Wisconsin.

Lao Tzu—first philosopher of Chinese Taoism, author of the Tao Te Ching.

League of Iroquois Nations—see Iroquois League.

Lewis and Clark Expedition—occurred 1804–1806, the first recorded overland expedition by any European from the Atlantic coast of North America to the Pacific Coast and back. It was conducted under the leadership of Meriwether Lewis and William Clark.

Light Circle—an association of beings working together, usually pertaining to an angelic association, but sometimes pertaining to human associations when angels incarnate.

Mayan—native American race in southern Mexico and Central America whose civilization dominated the area from approximately 550 B.C. to 1400 A.D. Alternate and less frequent usage uses the term to refer to the angelic race who influenced the Mayan people.

Mayan Calendar—ancient timetable carved in stone by the Mayan inhabitants of the Yucatan peninsula about 500 B.C. Detailed analysis of this calendar appears in art historian Jose Arguelles' book *The Mayan Factor*.

Medicine Wheel—complete cycle of healthy life within any system: universal, solar, terrestrial, bioregional, social or personal. The law of giving and receiving that makes for such healthy life. The sum of native American teachings. Also referred to as the Sacred Hoop and the Sacred Circle.

Mohawk—People of the Flint Stone. Iroquoian-speaking North American tribe occupying an area west of what is now Schenectady, New York, one of the original five nations to compose the Iroquois Confederacy. Within the confederacy they were given the title, *The Keepers of the Eastern Door.* Also called the Ganienkehaga.

Mother of the Old Ones—one of many titles for God in her female aspect, see Great Spirit.

Mount Shasta—an impressive double-peaked, extinct volcano of the Cascade Range in northern California, held sacred by native American tribes, as well as by many of the present inhabitants of the region.

Nagual—infinite sea of being that surrounds the universe like the sea surrounds a fish; contains the patterns of all possible new creation, the realm of potential and infinite energy; the non-dimensional Presence of God; that which has no name. Judeo-Christian: God the Father.

Native American—term for the aboriginal inhabitants of the Americas, increasingly preferred over the misnomer "Indians." Also used to apply to anything originating in America.

New Order of the Ages—words written on the Great Seal of the United States of America, pertaining to and anticipating the age that is now dawning.

New Sacred Reality—state of unity, new age, kingdom of heaven.

Nordic—of or relating to peoples from the northern countries of Europe.

Olmec—a Pre-Columbian civilization that originated in the lowlands of southern Veracruz and Tabasco in Mexico, considered some of the greatest sculptors of Meso-America.

Omaha—North American Plains people of the Siouan language stock, migrated westward from the Atlantic coast to Nebraska.

Oneida—People of the Standing Stones. One of the five founding tribes of the Iroquois Confederacy.

Ongwhehonwhe—[ɔ̃ŋ gwe ɔ̃:weh] the people true to reality, the real people, a term primarily used by the Iroquois but technically pertaining to (and used in this book to refer to) all people in whom spirit is fully incarnate and in whom spirit/ego integration has occurred. The Real People. The Original Beings. The Original Creators of Life.

Onondaga—People of the Hills. One of the five founding tribes of the Iroquois Confederacy.

Onondaga Lake—lake in upper New York State near the present city of Syracuse. Site of the founding of the League and planting of the Great Tree of Peace.

Order of the Ages—see New Order of the Ages.

Osage—North American Siouan-speaking tribe. The Osage migrated westward from the Atlantic coast and eventually settled in the Ozark Plateau and the prairies of what is now western Missouri.

Parliament—legislative assembly of Great Britain, consisting of the House of Lords and the House of Commons.

225

People of the Flint Stone—see Mohawk.

People of the Great Hills—see Seneca.

People of the Hills—see Onondaga.

People of the Plains—native American inhabitants of the area between the Mississippi River and the Rocky Mountains, including portions of the United States and Canada.

People of the Standing Stones—see Oneida.

People of the Swamps—see Cayuga.

Pilgrims—settlers of Plymouth, Massachusetts, the first permanent European colony in New England in 1620, separatists from the Church of England, emigrated first to the Netherlands and then to North America from Plymouth, England.

Pre-Columbian—term sometimes used to describe the period of time in North and South America before Christopher Columbus opened the two continents to European migration and trade in 1492.

Quetzalcoatl—[Kēt săl′kō-átl] a Bird Tribe incarnation in Central America, later considered a god of the Aztec and Toltec people, represented as a plumed serpent.

Real People—people of reality, people true to reality, people living in union with their spirits, see Ongwhehonwhe.

Rebellion, the—see Fall.

Sacagawea—[Sak′a-ǰa-wi′a?] a young Shoshoni woman from the Rocky Mountains who guided Lewis and Clark to the Pacific Ocean in 1806.

Sachem—a title sometimes given to the elected representatives of the Iroquois Confederacy; however, not a proper Iroquois term.

Sacred Circle—see Medicine Wheel.

Sacred Hoop—see Medicine Wheel.

Sacred Way—a way of balanced living in harmony with the Great Spirit and with all life. The Way of the Medicine Wheel. Also, the Way of Balance, see Medicine Wheel. Chinese: the Tao.

Second Self—subjective, body-associated sense of individual self. See ego.

Seneca—people of the Great Hills; one of the original five founding tribes of the Iroquois Confederacy, given the title within the Confederacy of *The Keepers of the Western Door.*

Sioux—a native American plains people. Also one of the six distinct language families or stocks that were represented in the plains area. The Siouan language stock included the Mandan, Hidatsa, Crow, Dakota-Assiniboin, Omaha-Ponca-Osage-Kansa, and Iowa-Oto-Missouri.

Six Nations—the Iroquois Confederacy after 1722, including the Tuscarora as well as the Mohawk, Oneida, Onondaga, Cayuga and Seneca.

Solar Clan—angels, spirits of the stars, see Bird Tribes.

Standing Hollow Horn—chief of the Sioux at the time of White Buffalo Calf Woman's historic visit.

Sun God—solar deity, the Being whose physical form is the sun, one of two primary aspects of the Great Spirit that with the earth provides the physical basis for biological life in this solar system.

Sword of Truth—symbolism for the nature of truth; the sword divides the real from the unreal, it distinguishes illusion from reality.

Tadodaho—[Ta'do da:hoh] a form of Adadarhoh, see Adadarhoh.

Tao Te Ching—classic of Chinese literature, outlining in simple and powerful verse a way of life in harmony with God and nature, a Chinese articulation of the Sacred Way of the Medicine Wheel.

Thunderbird—a rarely used name for the Great Spirit, also thunder, lightning and rain personified as a huge bird. See Great Spirit.

Thunder Tribes—angels. Winged Ones. See Bird Tribes.

Toltec—Nahuatl-speaking people who dominated what is now central Mexico from the tenth to the twelfth century.

Tonal—the universe and all it contains. Existing material structure. Every nameable thing. Universal Mother.

Tree of Peace—symbolic representation of peace. Also a literal tree planted on the shores of Lake Onondaga at the founding of the Iroquois Confederacy.

Tree of the Great Long Leaves—another name for the Tree of Peace.

Tree of Understanding—symbolic representation of understanding. Used interchangeably with Tree of Peace.

Tree of War—symbolic representation of warfare. The opposite of the Tree of Peace. On two occasions Deganawida used literal trees to represent the symbolic Tree of War.

True Self—eternal spirit self that proceeds from and knows God. Higher self. See Great Spirit.

True Human—the human in whom spirit-ego integration has occurred. One who is fully incarnate. A human being who embodies the balance between time and eternity, Creator and creation, earth and sun, spirit and matter. See Ongwhehonwhe.

Tuatha de Danann—aboriginal, pre-Celtic inhabitants of Ireland, one of the last of European peoples in whom the Bird Tribes were active and incarnate.

Tuscarora—an Iroquois tribe that joined what remained of the Confederacy in 1722, giving it the name of the Six Nation Confederacy.

Ulikron—a Bird Tribe incarnation among the Guaymi tribe of what is today Panama, also called The Golden One of the Sun.

United States system of government—system of federal government in which certain powers are delegated to the national government and all other powers fall to the states. The national government consists of executive, legislative and judicial branches, designed to check and balance each other. This system is supported by a Bill of Rights, which guarantees "inalienable" rights to all U.S. citizens.

Universal Mother—the Great Being whose body is the universe. Tonal. See Great Spirit.

Wakan Tanka—another name for God. See Great Spirit.

Wambli Galeshka—the spotted eagle.

Warrior tribes—human tribes given to warfare and fear-centered thought.

White Buffalo Calf Woman—an incarnation of the Great Spirit who taught the people of the Sioux the ceremony of the Sacred Pipe.

White Tail—a species of deer prevalent in Iroquois territory.

Winged Ones—angels, spirit beings. See Bird Tribes.

Yucatan—a peninsula mostly in southeast Mexico separating the Caribbean from the Gulf of Mexico. Center of Mayan civilization.

Yukon—a major river running from the Yukon territory of northwestern Canada westward through Alaska and into the Pacific ocean.

NOTE: The phonetic transcriptions are done according to the International Phonetic Alphabet with the addition of the following: ɔŋ pronounced as õnh; ẽ pronounced as ẽnh, : = length.